Moral Responsibility and Risk in Society

Risks, including health and technological, attract a lot of attention in modern societies, from individuals as well as policy-makers. Human beings have always had to deal with dangers, but contemporary societies conceptualise these dangers as risks, indicating that they are to some extent controllable and calculable. Conceiving of dangers in this way implies a need to analyse how we hold people responsible for risks and how we can and should take responsibility for risks.

Moral Responsibility and Risk in Society combines philosophical discussion of different concepts and notions of responsibility with context-specific applications in the areas of health, technology and environment. The book consists of two parts addressing two crucial aspects of risks and responsibility: holding agents responsible, i.e. ascribing and distributing responsibility for risks, and taking responsibility for risk. More specifically, the book discusses the values of fairness and efficacy in responsibility distributions and makes distinctions between backward-looking and forward-looking responsibility as well as individual and collective responsibility. Additionally, it analyses what it means to take responsibility for technological risks, conceptualising this kind of responsibility as a virtue, and furthermore, explores the notion of responsible risk communication and the implications for adult–child relationships.

This book will be of great interest to students and scholars of environmental ethics, bioethics, public health ethics, engineering ethics, philosophy of risk and moral philosophy.

Jessica Nihlén Fahlquist has a Ph.D. in philosophy and is a senior lecturer in biomedical ethics at the Centre for Research Ethics and Bioethics at Uppsala University in Sweden.

Earthscan Risk in Society Series
Edited by Ragnar E. Löfstedt
King's College London, UK

The Spatial Dimension of Risk
How Geography Shapes the Emergence of Riskscapes
Detlef Müller-Mahn

Effective Risk Communication
Edited by Joseph Arvai and Louie Rivers

Protecting Seniors Against Environmental Disasters
From Hazards and Vulnerability to Prevention and Resilience
Michael R. Greenberg

Anthropology and Risk
Åsa Boholm

Explaining Risk Analysis
Protecting health and the environment
Michael R. Greenberg

Risk Conundrums
Solving Unsolvable Problems
Edited by Roger E. Kasperson

Siting Noxious Facilities
Integrating Location Economics and Risk Analysis to Protect Environmental Health and Investments
Michael R. Greenberg

Moral Responsibility and Risk in Society
Examples from Emerging Technologies, Public Health and Environment
Jessica Nihlén Fahlquist

For more information about this series, please visit: www.routledge.com/Earthscan-Risk-in-Society/book-series/ERSS

"Jessica Nihlén Fahlquist offers a clear, analytic discussion of moral problems that arise as we seek to hold people responsible for the risks (rather than the actual harm) they impose on us and others. Her discussion of fair risk distributions, as well as the distinction between backward and forward-looking responsibility, will be of significant interest to anyone working in the field. The discussion of individual and collective responsibility adds valuable perspectives to an important academic debate with major societal implications. I warmly recommend this book to scholars, graduate students and others interested in the philosophy of risk and responsibility."

Martin Peterson, Professor of History and Ethics of Professional Engineering, Texas A&M University, USA

"Stressing both fairness and effectiveness, Jessica Nihlén Fahlquist builds a powerful case for action by governments and corporations to address systematic risks – from everyday risks like traffic safety to the enormous risks posed by greenhouse gas emissions and climate destabilisation. Nihlén Fahlquist's account is especially valuable because it distinguishes questions of blame from the challenge of deciding who should act now to deal with risks. Questions about 'backward looking responsibility' sometimes need to be asked. But to face up to risks, we need to look forward and find constructive ways to avoid, minimise and mitigate them. Using a wide range of examples, Nihlén Fahlquist argues eloquently that this usually involves collective responsibilities. And she shows that these collective tasks do not detract from individual responsibility. Instead, they enable people to make meaningful contributions – whether we think of safer driving or reducing environmental impacts or making responsible use of new technologies. Nihlén Fahlquist's writing is clear and calm, yet always keeps in sight the urgency of many risks created by modern societies and economies."

Garrath Williams, Senior Lecturer in Philosophy, Lancaster University, UK

Moral Responsibility and Risk in Society

Examples from Emerging Technologies, Public Health and Environment

Jessica Nihlén Fahlquist

Routledge
Taylor & Francis Group
LONDON AND NEW YORK

earthscan
from Routledge

First published 2019
by Routledge
2 Park Square, Milton Park, Abingdon, Oxon OX14 4RN

and by Routledge
52 Vanderbilt Avenue, New York, NY 10017, USA

First issued in paperback 2020

Routledge is an imprint of the Taylor & Francis Group, an informa business

British Library Cataloguing-in-Publication Data
A catalogue record for this book is available from the British Library

Library of Congress Cataloging-in-Publication Data
A catalog record has been requested for this book

ISBN 13: 978-0-367-58606-5 (pbk)
ISBN 13: 978-1-138-19290-4 (hbk)

Typeset in Sabon
by Wearset Ltd, Boldon, Tyne and Wear

For William, Isabella and Jonathan

Contents

Acknowledgements

There are a number of people in my life and at work without whom this book would never have been written. In terms of intellectual stimulation, the ideas in this book have been developed within the context of a never-ending discussion with my dear friend Sabine Roeser. Without her encouragement, inspirational attitude and friendship, I doubt that I would have dared to put all my disparate thoughts into one book. Sabine has inspired me beyond explanation, as a daring philosopher who has the ability to enthuse anyone to start, or continue, to do philosophy. In addition, I am grateful to my friends and colleagues at the Centre for Research Ethics and Bioethics at Uppsala University. It is a joy to work at such an inspirational and friendly interdisciplinary research centre.

Many of the ideas presented in this book have been affected by continuous discussions with my beloved husband, Rickard Fahlquist, who always provides insightful comments and questions as an engineer and an individual thoroughly engaged in society and humanity. I am intrigued by the fact that it is always fascinating to talk to him even though we have been talking for 22 years. My father, Gunnar Nihlén, a pragmatic engineer, who is deeply engaged in issues concerning sustainability, has also provided me with insightful comments throughout the years. Having two engineers close at hand forces a philosopher to constantly test her ideas. The encouragement from my father's wife, Birgitta Nihlén, my brother, Tomas Nihlén, and his partner, Linda Pierre, as well as from my parents-in-law, Eva and Per Fahlquist, has also been very important.

One person who has been essential in my life and work is my mother, Ylva Skogström, who died in the process of my finishing this book. How I would have wanted her to see my book and to read it! I cannot even imagine where I would have been without her endless love and support. I miss her deeply. I am endlessly grateful to my three wonderful children to whom I dedicate this book. I cannot imagine how boring life would be without them, not least as a contrast to academic life. In addition to my endless love for these three very special individuals, they provide me with a constant challenge to explain what I do and why it is important. Their questions are also intellectually stimulating as they force me to try to be clear and concise. It surprises me again and again how sound the moral intuitions of children are once we take the time to listen to them. Some children are truly more virtuous than many adults. I am so grateful that I have the opportunity to live close to three of them and experience their moral development.

I would also like to direct a special thanks to my co-authors: Sabine Roeser, Ibo van de Poel and Kalle Grill, for helping me develop my ideas, for being excellent collaborators and for agreeing to my reuse of our, to some extent, common work. Last, but not least, I am deeply grateful to Michele Rosen for reading and commenting on my chapters and helping me clarify vague statements and improve my language. Anything that is still not clear or adequately readable is of course my own responsibility.

Permissions

The chapters in this book are, more or less, reprinted or based on previously published articles. The articles, or parts of them, are reprinted with the kind permission of the publishers. Furthermore, in relevant cases, co-authors have agreed to the reprinting of parts of the published articles. In those cases, those parts for which Jessica Nihlén Fahlquist was mainly responsible are reprinted in this book. Nevertheless, some of the material was originally co-authored. The articles on which the chapters are based are the following:

Introduction to Part I
Nihlén Fahlquist, J. 2008. Moral responsibility and the ethics of traffic safety. Royal Institute of Technology: Stockholm, PhD thesis.

Chapter 1
Nihlén Fahlquist, J. 2006. Responsibility ascriptions and public health problems. Who is responsible for obesity and lung cancer? *Journal of Public Health* 14(1): 15–19.

Chapter 2
Nihlén Fahlquist, J. 2006. Responsibility ascriptions and Vision Zero. *Accident Analysis and Prevention* 38: 1113–1118.

Grill, K. and Nihlén Fahlquist, J. 2012. Responsibility, paternalism and alcohol interlocks. *Public Health Ethics* 5(2): 116–127.

Chapter 3
Nihlén Fahlquist, J. 2009. Moral responsibility for environmental problems – individual or institutional? *Journal of Agricultural and Environmental Ethics* 22(2): 109–124.

Introduction to Part II
Nihlén Fahlquist, J. 2015. Responsibility as a virtue. In van de Poel, I., Lambér Royakkers, L. and Zwart, S. (eds), *Moral Responsibility and the Problem of Many Hands*. London: Routledge.

Chapter 4
Nihlén Fahlquist, J. 2017. Responsibility analysis. In S.O. Hansson (ed.), *Ethics of Technology: Methods and Approaches*. London: Rowman & Littlefield.

Nihlén Fahlquist, J. 2015. Responsibility as a virtue. In van de Poel, I., Lambér Royakkers, L. and Zwart, S. (eds), *Moral Responsibility and the Problem of Many Hands*. London: Routledge.

Chapter 5
Nihlén Fahlquist, J. and Roeser, S. 2015. Nuclear energy, responsible risk communication and moral emotions: a three level framework. *Journal of Risk Research* 18(3): 333–346.

Nihlén Fahlquist, J. 2018. Vaccine hesitancy and trust. Ethical aspects of risk communication. *Scandinavian Journal of Public Health* 46(2): 182–188.

Nihlén Fahlquist, J. 2016. Experience of non-breastfeeding mothers, norms and ethically responsible risk communication. *Nursing Ethics* 23(2): 231–241.

Nihlén Fahlquist, J. and Roeser, S. 2011. Ethical problems with information on infant feeding in developed countries. *Public Health Ethics* 4(2): 192–202.

Chapter 6
Nihlén Fahlquist J. and van de Poel, I. 2012. Technology and parental responsibility – the case of the V-chip. *Science and Engineering Ethics* 18: 285–300.

Nihlén Fahlquist, J. 2015. Responsibility and privacy – ethical aspects of using GPS to track children. *Children & Society* 29(1): 38–47.

Introduction

Moral responsibility and risk

Risks are assessed, managed and discussed on an everyday basis in modern society. The interdisciplinary study of risk that exists today arose when the need to regulate new technology was introduced by the era of industrialism. Through development and innovation in technology and medicine, societies as well as people have gained a number of benefits related to health, well-being, transportation and economy. However, these new developments also entail risks. Formal risk analysis was introduced during the first part of the twentieth century, primarily by engineers, epidemiologists, actuaries and industrial hygienists (Krimsky and Golding, 1992). The public started to pay attention to the environmental and health risks that were created alongside the benefits of technological innovation and development during the latter part of the twentieth century. Public awareness of these risks emerging in the 1960s led to a new multidisciplinary field of research: risk analysis and risk research (Hansson, 2009). The new field of research included aspects that had not been parts of the formal risk analysis done previously. In this context, social scientists started to focus on risk perception. It was acknowledged that experts and lay people have very different ideas about what an acceptable risk is, and risk communication was seen as a tool to bridge the gap and solve "the public acceptance problem" (e.g. Slovic, Fischoff and Lichtenstein, 1976; Krimsky and Golding, 1992). The public has generally been seen as "irrational" and "emotional", with regard, for example, to fears about nuclear energy, when experts calculate those risks as "negligible" (Roeser, 2011; Nihlén Fahlquist and Roeser, 2015). In contemporary debate and in the context of health issues, we see the same thing happening with, for example, vaccination policy (Nihlén Fahlquist, 2018). People who are "overly" worried are seen as anti-science, irrational, or even immoral or lacking in social solidarity. In this book, we will examine these and other examples and explore arguments showing that the picture is far more complex than this dichotomisation of experts and lay people suggests.

The philosophy and ethics of risk emerged in this context. As noted by Hansson, the risky and uncertain nature of human life was the "blind spot" of traditional moral philosophy, dealing as it did with clearly delineated actions with well-defined consequences (Hansson, 2009). In contrast, the ethics of risk studies the ethical aspects of situations in which the outcomes of possible courses of action are not known (e.g. Altham, 1983; Asveld and Roeser, 2009; Hansson, 2013; Roeser *et al.*, 2012). This is more realistic, because a significant share of our collective activities in contemporary society deals with questions relating to the analysis, management and communication of health risks, environmental risks and political and economic risks. Hence, it is reasonable that ethical theories should engage with issues of risk. However, this is not as straightforward as one might think. Ethical theories traditionally deal with actions where the outcome is certain (Hansson, 2009). When applying ethical theories such as the utilitarian framework, actions should be evaluated as right or wrong because of their consequences. Alternatively, an action can be considered right or wrong because it does or does not conform to deontological rules, duties and rights. Asking such questions presupposes that we know what will happen if we do X or Y. The rightness or wrongness of an act is evaluated according to the outcome or intention in the context of different states of affairs. However, in contemporary society the outcomes of a significant share of our activities, particularly those that may affect other human beings or the environment, are not at all certain; they might, may or could lead to X, Y or Z, or to something completely different. We are not dealing with a situation in which we can choose one of two options, knowing what they will lead to. We drive our cars to work in the morning, an everyday activity that *could* nevertheless contribute to our grandchildren and their generation not being able to live in certain places or being exposed to natural disasters such as flooding. At the same time, that particular instance of driving does not really lead to the flooding; rather, it is all the individual instances of driving taken together, our collective action of extensive driving, if you will, that does increase the probability that our grandchildren will have environmental problems, to say nothing of the fact that people in some areas of the world already have problems. And yet we do not intend to harm anyone when taking our car to the grocery store.

Similarly, when I am driving, I might end up looking at my mobile phone in an unfortunate moment and thereby tragically cause someone's death. When I allow or compel my child to receive a new pandemic vaccine, I may contribute to the protection of vulnerable people who cannot receive the vaccine. However, I might also expose my child to the risk of developing diseases that the tests conducted so far on the vaccine have yet not discovered. Our actions in modern society are intertwined, globally connected and complex. How do we know what is the right thing to do in such a complex context? This is the question that I attempt to explore. Even though the answer is not easily accessible, it hopefully benefits from a conceptual and normative analysis to which I hope to contribute.

The thesis of this book

This book aims to contribute to the ethics of risk by exploring issues concerned with moral responsibility and risk. The underlying thesis of this book is that the most promising way forward requires that we stop approaching questions of risk from the perspective of ethical theories, asking what the right or wrong decision or policy is. Instead, problems of risk should be approached through the concept, or rather concept*s*, of responsibility. As we will see, there are clear conceptual connections between risk and responsibility because risks in modern society are conceptualised as something that could, and should, be controlled and managed. This notion of control explicitly, or sometimes implicitly, means that someone is to *be held responsible* or should *take* responsibility for risks. If human beings in premodern times conceived of dangers and threats as stemming from God or Fate, risks in contemporary society are conceived as something for which someone, an individual or a collective, is responsible. To approach risk problems from the perspective of moral responsibility entails not endorsing one specific theory, while at the same time gathering risk issues under one conceptual umbrella. However, as we will see, the notions of moral responsibility are varied and nuanced enough to capture several dimensions of risk-related societal challenges. The main distinction to be made, according to which the book is organised, is the one between responsibility as something external, that we ascribe and distribute, and the more internal notion of actively taking responsibility. Simply put, we *hold* individuals and organisations responsible, but we can and should also *take* responsibility for risks. The question is under what conditions and circumstances the two general notions are relevant. The chapters in Part I focus on the notion of holding agents responsible, i.e. ascribing and distributing responsibility. Part II focuses on the notion of taking responsibility.

Thus, this book focuses on ethical aspects of risk, particularly related to different notions of moral responsibility. In order to be able to understand the arguments and discussions, we need to start with a discussion of the core concepts: moral responsibility and risk. These concepts and the philosophical discussion focusing on them will be explored in the introduction to Part I. In addition to the introduction, Part I consists of three chapters focusing on three different aspects, or dimensions, of ascriptions and distributions of responsibility. In Chapter 1, it is argued that responsibility ascriptions and distributions can be analysed in terms of two values: fairness and efficacy. In Chapter 2, the distinction between backward-looking and forward-looking responsibility is discussed. In Chapter 3, aspects related to individual and collective responsibility are introduced.

Whereas Part I focuses on responsibility ascriptions and distributions and the ethical aspects of our holding agents responsible, Part II focuses on the notion of taking responsibility or being a responsible person, conceptualising it in terms of a virtue. Chapter 4 explores how ethical technological development needs this kind of responsibility-taking to reduce and manage risks. Chapter 5 discusses responsible risk communication as requiring an ethically acceptable message, a legitimate procedure and an ethical analysis of its effects. When conceiving of responsibility as a virtue, as something to be developed over a lifetime with the help of role models and through experience, children come to the fore. For this reason, the concluding chapter, Chapter 6, deals with children as developing moral agents and how their development is facilitated or impaired by adults and through technology. Throughout the book, examples are provided from emerging technologies, public health and environment.

References

Altham, J.E.J. 1983. Ethics of risk. *Proceedings of the Aristotelian Society* 84: 15–29.

Asveld, L. and Roeser, S. (eds). 2009. *The Ethics of Technological Risk*. London: Earthscan.

Hansson, S.O. 2009. An agenda for the ethics of risk. In L. Asveld and S. Roeser (eds), *The Ethics of Technological Risk*. London: Earthscan.

Hansson, S.O. 2013. *The Ethics of Risk: Ethical Analysis in an Uncertain World*. Basingstoke: Palgrave Macmillan

Krimsky, S. and Golding, D. 1992. *Social Theories of Risk*. Westport, CT: Praeger.

Nihlén Fahlquist, J. 2018. Vaccine hesitancy and trust: ethical aspects of risk communication. *Scandinavian Journal of Public Health* 46(2): 182–188.

Nihlén Fahlquist, J. and Roeser, S. 2015. Nuclear energy, responsible risk communication and moral emotions: a three level framework. *Journal of Risk Research* 18(3): 333–346.

Roeser, S. 2011. Nuclear energy, risk, and emotions. *Philosophy and Technology* 24(2): 197–201.

Roeser, S., Hillerbrand, R., Sandin, P. and Peterson, M. (eds). 2012. *Handbook of Risk Theory: Epistemology, Decision Theory, Ethics, and Social Implications of Risk*. Dordrecht: Springer.

Slovic, P., Fischoff, B. and Lichtenstein, S. 1976. Cognitive process and societal risk taking. In J.S. Carroll and J.W. Payne, *Cognition and Social Behaviour*. Hillsdale, NJ: Lawrence Erlbaum.

Part I

Holding agents responsible for risk

Introduction to Part I

Moral responsibility – the philosophical discussion

Interestingly, the word "responsibility" is fairly new to the history of philosophy. The oldest confirmed use of the term in English is from 1642, but it existed in French in the fifteenth century (Oxford English Dictionary). Responsibility became an important concept in a political context when representative governments emerged in Europe towards the end of the eighteenth century (McKeon, 1957: Williams, 2008).

The novelty of the term does not mean that we did not hold each other accountable prior to modern society. Aristotle, who is often mentioned in philosophical texts about responsibility, wrote primarily about responsibility, without using this term, as blameworthiness. Aristotle states that if an agent did something wrong voluntarily, i.e. freely and knowingly, he is blameworthy (Aristotle, 2000).

In the philosophical literature, the most contested question concerning responsibility is whether an agent can reasonably be held responsible for her actions even if there is no such thing as free will. If it turns out that human beings are predetermined to act in certain ways, i.e. that we have no choice, are we morally responsible? In essence, is moral responsibility compatible with causal determinism? This debate on moral responsibility is carried out within the frameworks of compatibilism and incompatibilism, the former arguing that responsibility is compatible with determinism and the latter denying this (e.g. Fischer and Ravizza, 1993). Incompatibilists argue that if we do not have free will, we are not responsible for our actions. In contrast, compatibilists find different ways to argue that we can still reasonably ascribe responsibility to individuals.

The question of free will and its potential compatibility with moral responsibility may at times seem very esoteric. However, it is increasingly relevant due to neuroscientific research. For example, evidence suggests that the brain dysfunction suffered by psychopaths prevents them from having adequate cognitive-affective control over motivational states, which would arguably make it unreasonable to blame them for wrongdoing, or at least harm-doing (Glannon, 2008).

This question is also relevant in relation to genetic research. The media regularly reports about the discovery of new genetic connections. In some cases, it is merely the case of a gene supposedly associated or linked to a certain medical condition, but in other cases researchers claim to have found a genetic association for a certain type of behaviour, e.g. antisocial behaviour in criminals (Tiihonen *et al.*, 2015). Some geneticists point out that such studies should be treated with caution due to the uncertainty regarding clinical validity (e.g. Howard, Avard and Borry, 2011; Janssens *et al.*, 2008). However, the more research is done in this field, the more relevant questions concerning the extent of free will and, correspondingly, moral responsibility become.

The concept of responsibility in the discussion of free will and determinism focuses on the relationship of moral agency and capacity to responsibility as blameworthiness. It is a largely backward-looking use of the concept of moral responsibility. In a sense, the participants in this debate are interested in reasonable conditions for holding someone responsible in the sense of blameworthiness and praiseworthiness. If we do not have free will, but are totally determined by genetic setup and upbringing, is it really reasonable to blame and praise us for our actions?

In ordinary language, as well as in the philosophical debate, a basic notion is that a person is responsible if certain conditions are fulfilled. For example, if someone was forced to do something that we normally consider wrong, we are generally reluctant to hold her responsible, i.e. we would excuse her for acting wrongfully. The following is a useful list of conditions that are often seen as prerequisites for holding an agent morally responsible in the sense of blameworthiness (Van de Poel *et al.*, 2012):

1 Capacity.
2 Causality.
3 Knowledge.
4 Freedom.
5 Wrongdoing.

We will have the opportunity to analyse the conditions and the distinction between backward-looking and forward-looking responsibility in Chapter 2. At this point, we should simply acknowledge that this philosophical discussion concerning the reasonableness of ascribing responsibility at all focuses on moral agency. I will now introduce an additional complexity of moral responsibility, namely the presence of affective as well as cognitive aspects in responsibility ascriptions.

Emotions and reflection

Let us imagine that an innocent person is stabbed to death. A common first reaction for most people when someone is stabbed is emotional. We would probably get angry at the perpetrator and feel sympathy for the victim and her family. Perhaps we think about what we would do if it happened to a family member or friend. With time, the emotional reaction is accompanied, and perhaps moderated, by reflection and analysis. We may find out that the perpetrator committed the crime after having been released from the psychiatric clinic where he had spent most of his adult life due to regular psychoses. Questions that may come to mind are: What caused the killer to stab an innocent person? What was the extent of the perpetrator's psychiatric problems? Could social services, hospitals, or other institutions have acted differently to prevent the tragedy? If they could and we decide to hold others responsible in addition to the main perpetrator, does this necessarily reduce or eliminate his responsibility? Questions such as these are merely a few of the questions that contribute to making our thinking about the event more complex and detached than our initial emotional reaction (Nihlén Fahlquist, 2008).

Traditionally, philosophical thought distinguishes between reason and emotion, and the former has been seen as better for decision-making. In psychology, the so-called dual process theory states that human beings use two systems when making decisions. System 1 is emotional and spontaneous; this system helped human beings when there were threats in the form of lions and tigers. In contrast, System 2 is rational-analytical, slower, and emerged later in human evolution. System 2 is presumably preferable because it helps us make rational decisions (e.g. Kahneman, 2003; Roeser, 2012; Slovic, 2002; Slovic *et al.*, 2005). Furthermore, one of the central questions in metaethics and empirical moral psychology has been whether moral judgements are a consequence of reason or emotion. It has often been assumed that reason and emotion are independent and separate. However, there is now a growing body of research indicating that they are not as independent and distinguishable as previously thought (e.g. Saunders, 2015; Roeser, 2017). Studies even indicate that we *need* emotions in order to make rational decisions (Damasio, 1994).

If we think about emotional reactions in the case of the aforementioned stabbing, dual process theory and the distinction between instant emotions and slower reflection appears to fit the way we might analyse the situation. Yet, as described above, research now indicates it is not as simple. Interestingly, one philosophical idea that has been very influential equates responsibility with the "reactive attitudes" that we have as human beings to the actions of other members of our moral community. In his famous article from 1962, "Freedom and resentment" (Strawson, 1962), Strawson asserted that moral responsibility is merely part of our social and political practice and does not depend on any metaphysical idea about freedom. Whether we hold someone responsible does not depend on whether she had free will. The reactive attitudes, e.g. resentment and moral indignation, are all there is to moral responsibility. It is our expectations of our own and others' behaviour that lead to feelings of guilt, remorse and responsibility, according to this theory. When we have been injured or offended by someone else's actions, we are normally expected to feel resentment unless there are special considerations. These considerations are very similar to the conditions mentioned above and relate to the status of the perpetrator as moral agent or to his lack of knowledge. When such special considerations are relevant, our reactive attitudes are modified accordingly (Strawson, 1962).

However, our reactions, or reactive attitudes, do not to tell the whole story of moral responsibility. As members of the moral community, we also have the ability to step back and analyse the situation from a more detached perspective. So, even if the dual process theory and the traditionally assumed dichotomy between reason and emotion is not valid, there is something to be said for reflection or deliberation in contrast to immediate reaction. This does not mean that reflection or deliberation is unemotional or non-emotional. Reflection might very well be based on a combination of cognition and affect (cf. Roeser, 2017). Roeser discusses this idea in relation to technological risk, arguing that responsible engineers ideally use their moral emotions in order to make good decisions concerning risk. Similarly, the moral emotions of laypeople should be taken seriously when deliberating about the ethics of risk. It is important to note that moral emotions are different from gut reactions and are instead based on deliberation (Roeser, 2006; Roeser, 2017).

If we consider these insights, which show that there is a difference between immediate reaction or gut reaction and the slower deliberative thinking that could be based on emotions as well as reasons, we can see that there is something lacking in the notion of responsibility *as* reactive attitudes. There is a difference between being *held* responsible and *being* responsible that is not captured by this notion of responsibility as merely a matter of reactive attitudes. We are able to, and regularly do, criticise and question our own and others' reactive attitudes. Hence, to be morally responsible is not merely to be a target of reactive attitudes. According to Fischer and Ravizza's revised Strawsonian theory, "agents are morally responsible if and only if they are *appropriate* recipients of reactive attitudes" (Fischer and Ravizza, 1993; Nihlén Fahlquist, 2008). Hence, whereas the first element of moral responsibility involves the reactive attitudes towards the actions of co-agents in a moral community, the other element is critical reflection about these practices. It should be kept in mind that we are still talking about responsibility primarily as blameworthiness. Although not the only notion of responsibility, it is an important one and it is applicable to contexts such as engineering. When risk events occur, or warning signs are ignored, the question arises as to whether someone is to blame. If it turns out that an engineer, for example, did something wrong, she may be blamed. However, if we also get the information that she was not provided with correct information or that her manager forced her to act wrongfully, our willingness to hold her responsible is probably modified.

Discussions about societal problems, for example the debate about who is responsible for the public health problem related to obesity, are multilayered and complex. At the individual level, the question is whether an agent, A, is responsible for becoming obese and for the health-related consequences of that. To evaluate her responsibility, we would want to know A's circumstances and whether she has any valid excuse. Was she, for instance, aware of the risks of obesity? Did her parents teach her how to develop a healthy lifestyle? Furthermore, as I explore in Chapter 3, what alternatives did she have? If she could not afford to exercise at a gym or buy healthy food, this may reduce her responsibility. Additionally, there may be genetic factors that facilitate or impede weight loss.

At the collective level, the question is whether other actors, for instance, the government, authorities, the food industry, health care institutions and so forth, share responsibility with the individual. To estimate their role in A's and other people's obesity, we would be interested to know more about their causal contribution, whether they had any valid excuse for their actions, and the way in which they create and affect the alternatives that are open to individuals. At the individual as well as the collective level, we are interested both in what caused A and others to become obese and whether anyone should be blamed, but also how this problem should be solved and who should solve it. Conflating backward-looking and forward-looking responsibility can have negative consequences in moral as well as efficiency terms. Agents could be unjustifiably blamed and communication problems could hamper the process of solving the problem at hand.

Causation and responsibility

Acknowledging that responsibility is sometimes aimed at solutions in a more forward-looking way sheds light on the question of how important causation is to ascriptions of moral responsibility. A common assumption when discussing moral responsibility is that:

i If agent A caused state-of-affairs S, she is responsible for it.
ii If A is responsible for S, she caused it.

(Nihlén Fahlquist, 2008)

This strict relationship between causation and responsibility is either assumed by philosophers or, as the extensive literature on free will and determinism shows, the exact content of this kind of causation is the focus of the discussion. Causal responsibility is so closely attached to the concept of moral responsibility that it is sometimes difficult to detach the two. Intuitively, we are only morally responsible for what we cause (Sartorio, 2004).

As Fredriksen argues, the standard philosophical conception of responsibility, focusing on "responsibility because of control", is too narrow and should be supplemented with "responsibility because of social involvement" (Fredriksen, 2005). He claims that the standard model is wrong because we do not merely ascribe responsibility to agents who have had control over their actions. He gives an example in which a young father is told that he has a serious neurological disease that will cause his body to degenerate and will lead to his death within ten years. Obviously, neither the father nor his wife controlled or caused this in any way, but they still have a responsibility to see to it that their children are taken care of and provided for and so forth. He is not exempt from responsibility because he did not control or cause the problem.

Young has presented a similar conception of responsibility: the social connection model of responsibility. She examines the way in which individuals in the developed world are responsible for the conditions of workers in the global apparel industry located in third world countries, although she argues that her model is applicable to all cases of structural injustice. Young argues that "all agents who contribute by their actions to the structural processes that produce injustice have responsibilities to work to remedy these injustices" (Young, 2006). There are five distinguishing features of this concept:

- it does not isolate perpetrators.
- it evaluates the background conditions of action.
- it is more forward-looking than backward-looking.
- its responsibility is essentially shared.
- it can be discharged only through collective action.

This concept of responsibility focuses more on social connection than causation. Of course, causal links are not completely irrelevant since (a) the agents who are ascribed responsibility *contribute by their actions* to structural injustice and (b) forward-looking causal power is necessary to achieve the desired collective action. However, the social connection model focuses less on backward-looking causal responsibility as a necessary requirement for moral responsibility, and as such it could be a promising starting point for the development of the forward-looking concept of responsibility. We will explore the difference between backward-looking and forward-looking responsibility in Chapter 2.

Risk

Just as when we talk about responsibility, it is not always clear what we mean when we talk about risk. Sometimes we merely mean danger or threat, but at other times a stricter definition is assumed. Experts usually refer to the product of a probability and a negative outcome, or simply the probability that something bad will happen (cf. Asveld and Roeser, 2009; Hansson, 2009).

Risk ethicists argue that conventional risk analysis should include aspects not fully covered by the current methods. Experts usually characterise risks against the background of risk-benefit and cost-benefit analysis. However, lay people consider a wide range of other aspects, for example whether the risks and benefits of a certain activity are distributed fairly, whether the risk is voluntarily taken or imposed by others, and whether the risk is controllable (e.g. Slovic, 2000). These concerns are also normatively justified, as argued by philosophers (e.g. Roeser, 2007; Shrader-Frechette, 1991). Hence, not only *do* people care about fairness in a risk-benefit distribution, we *should* care about it (Roeser, 2006). If nuclear waste is located in areas where people are poor, but wealthy people enjoy the benefits, then this is an ethically questionable situation (Hermansson and Hansson, 2007). If people in the rich part of the world use pharmaceuticals but people in poor countries suffer the environmental consequences, then that is a problem of justice. Rethinking risk analysis along these lines could have substantial consequences for policy-making. According to conventional risk analysis it may be acceptable to, for example, direct heavy traffic through a certain residential area if the benefits in terms of transportation and cost savings outweigh the risks of environmental pollution and reduced traffic safety. If fairness is considered, however, the same infrastructural project may be judged ethically unacceptable.

Conceptual relations between risk and responsibility

The notions of risk and responsibility are closely interconnected. When we go from talking about a "danger" or "threat" to talking about a "risk", it is usually implicitly assumed that it is something that could, at least to some extent, be managed and controlled by human beings. In modern society, risk is usually seen as a calculable threat. Beck and Giddens famously describe modern society as a "Risk Society" and risks as introduced by modernisation (Beck 1992; Giddens 1999). Giddens argues that we live *after* nature and tradition:

> For hundreds of years, people worried about what nature could do to us – earthquakes, floods, plagues, bad harvests and so on. At a certain point, somewhere over the past fifty years or so, we stopped worrying so much about what nature could do to us, and we started worrying more about what we have done to nature.... To live after the end of tradition is essentially to be in a world where life is no longer lived as fate. For many people – and this is still a source of class division in modern societies – diverse aspects of life were established by tradition as fate.
>
> (Giddens, 1999, p. 3)

By calculating a number stating the probability that something negative will occur and the extent of the potential damage, the threat has moved from the realm of the unknown or uncertain to the realm of potential events that human beings can affect. The latter also implies that someone is, or will be, responsible for the consequences. If human beings can be part of the causal chain that increases or decreases the likelihood of a certain event, then someone could also be responsible for it. According to risk society theory, both risk and responsibility are connected to control and decisions. This implies a rather direct conceptual connection between risk and responsibility. As Giddens expresses it:

> The relation between risk and responsibility can be easily stated, at least on an abstract level. Risks only exist when there are decisions to be taken ... The notion of responsibility also presumes decisions. What brings into play the notion of responsibility is that someone takes a decision having discernible consequences.
>
> (Giddens, 1999, p. 8)

This idea of control, which is central to risk in the theory of risk society, is also central for moral responsibility (Van de Poel and Nihlén Fahlquist, 2012).

When a risk is identified, we explicitly or implicitly attach a notion of responsibility and action-taking to it. If we take an example from healthcare, in the fifteenth century fatal infections like measles were a serious threat worldwide. In 1912, the United States started to count the cases, declaring it a "nationally notifiable disease", which meant that the government would require healthcare providers and laboratories to report all diagnosed cases (CDCP, 2018). The danger had entered the realm of potentially humanly controllable risks. In 1963, the first measles vaccine, the so-called Edmonston-Enders strain, was licensed (CDCP, 2018).

In the cultural, religious and socio-economic context of the Middle Ages, an individual was considered lucky or unlucky and possibly affected by God's will and/or Fate if she got measles, which also determined whether she survived or died. It was an uncontrollable threat. In modern society there is a vaccine, and measles is considered a public health issue. It can perhaps not be completely controlled, but it can be managed and almost eliminated through the vaccine developed as part of a fundamental medical progress.

In the case of vaccination, questions that arise are: "Is having one's children vaccinated a parental responsibility?" Alternatively, "Is it the government's responsibility to make sure all children are vaccinated?" Some of these questions are context-bound, but many are perennial in all discussions on risk and responsibility. In Part I we will look at three clusters of perennial issues in discussions concerning ascriptions and distributions of responsibility for risk:

- Fairness versus efficacy.
- Backward-looking versus forward-looking responsibility.
- Individual versus collective responsibility.

In the example of vaccines, these ethical questions could, for example, be framed as follows:

- Would it be fair to ascribe most of the responsibility for having children vaccinated to the parents?
- Would it be more effective to make it the government's responsibility?
- If the incidence of measles starts to increase, whose fault is it?
- Who is most able to make necessary changes in order to promote vaccination?
- To what extent is it up to the individual to make sure a child and her family receive vaccinations, and to what extent is this a matter for legislation?

References

Aristotle. 2000. *Nicomachean Ethics* (edited and translated by R. Crisp). Cambridge: Cambridge University Press.

Asveld, L. and Roeser, S. (eds). 2009. *The Ethics of Technological Risk*. London: Earthscan.

Beck, U. 1992. *Risk Society: Towards a New Modernity*. London: Sage.

CDCP. 2018. www.cdc.gov/measles/about/history.html (accessed 7 August 2017).

Damasio, A. 1994. *Descartes' Error: Emotion, Reason, and the Human Brain*. New York: Putnam.

Fischer, J.M. and Ravizza, M. 1993. Introduction. In J.M. Fischer and M. Ravizza (eds), *Perspectives on Moral Responsibility*. Ithaca, NY: Cornell University Press.

Fredriksen, S. 2005. Luck, risk and blame. *Journal of Medicine and Philosophy* 30: 535–553.

Giddens, A. 1999. Risk and responsibility. *Modern Law Review* 62(1): 1–10.

Glannon, W. 2008. Moral responsibility and the psychopath. *Neuroethics* 1: 158–166.

Hansson, S.O. 2009. An agenda for the ethics of risk. In L. Asveld and S. Roeser (eds), *The Ethics of Technological Risk*. London: Earthscan.

Hermansson, H. and Hansson, S.O. 2007. A three-party model tool for ethical risk analysis. *Risk Management* 9(3): 129–144.

Howard, H.C., Avard, D. and Borry, P. 2011. Are the kids really all right? Direct-to-consumer genetic testing in children: are company policies clashing with professional norms? *European Journal of Human Genetics* 19(11): 1122–1126.

Janssens A.C., Gwinn, M., Bradley, L.A., Oostra, B.A., van Duijn, C.M. and Khoury, M.J. 2008. A critical appraisal of the scientific basis of commercial genomic profiles used to assess health risks and personalize health interventions. *American Journal of Human Genetics* 82: 593–599.

Kahneman, D. 2003. A perspective on judgment and choice: mapping bounded rationality. *American Psychologist* 58(9): 697–720.

McKeon, R. 1957. The development and the significance of the concept of responsibility. *Revue Internationale de Philosophie* 11(39): 3–32.

Nihlén Fahlquist, J. 2008. *Moral Responsibility and the Ethics of Traffic Safety*. Ph.D. thesis, Stockholm: Royal Institute of Technology.

Roeser, S. 2006. The role of emotions in judging the moral acceptability of risks. *Safety Science* 44: 689–700.

Roeser, S. 2007. Ethical intuitions about risks. *Safety Science Monitor* 11: 1–30.

Roeser, S. 2012. Emotional engineers: toward morally responsible design. *Science and Engineering Ethics* 18(1):103–115.

Roeser, S. 2017. *Risk, Technology, and Moral Emotions*. London: Routledge.

Sartorio, C. 2004. How to be responsible for something without causing it. *Philosophical Perspectives (Ethics)* 18(1): 315–336.

Saunders, L.F. 2016. Reason and emotion, not reason or emotion in moral judgment. *Philosophical Explorations* 19(3): 251–267.

Shrader-Frechette, K. 1991. *Risk and Rationality*. Berkeley, CA: University of California Press.

Slovic, P. 2000. *The Perception of Risk*. Earthscan: London.

Slovic P. 2002. The affect heuristic. In T. Gilovich, D. Griffin and D. Kahnemann, (eds), *Intuitive Judgment: Heuristics and Biases*. Cambridge: Cambridge University Press.

Slovic, P., Peters, E., Finucane, M.L. and MacGregor, D.G. 2005. Negative affect, risk, and decision making. *Health Psychology* 24: 35–40.

Strawson, P.F. 1962. Freedom and resentment. *Proceedings of the British Academy* 48: 1–25 (reprinted in J.M. Fischer and M. Ravizza (eds), *Perspectives on Moral Responsibility* (pp. 45–66). Ithaca, NY: Cornell University Press, 1993).

Tiihonen, J., Rautiainen, M.R., Olilla, H.M., Repo-Tiihonen, E., Virkkunen, M., Palotie, A., Pietiläinen, O., Kristiansson, K., Joukamaa, M., Lauerma, H., Saarela, J., Tyni, S., Vartiainen, H., Paananen, J., Goldman, D. and Paunio, T. 2015. Genetic background of extreme violent behaviour. *Molecular Psychiatry* 20: 786–792.

Van de Poel, I. and Nihlén Fahlquist J. 2012. Risk and responsibility. In S. Roeser, R. Hillerbrand, M. Peterson and P. Sandin (eds), *Handbook of Risk Theory: Epistemology, Decision Theory, Ethics, and Social Implications of Risk* (pp. 877–908). Dordrecht: Springer.

Van de Poel, I., Nihlén Fahlquist, J., Doorn, N., Zwart, S., Royakkers, L. and di Lima, T. 2012. The problem of many hands: climate change as an example. *Science and Engineering Ethics* 18(1): 49–67.

Williams, G. 2008. Responsibility as a virtue. *Ethical Theory and Moral Practice* 11: 455–470.

Young, I.M. 2006. Responsibility and global justice: a social connection model. *Social Philosophy and Policy* 23(1): 102–130.

1 Fairness and efficacy in responsibility distributions[1]

Introduction

The assignment of responsibility for risks often underlies societal discussions, albeit not always explicitly. When new research indicates, for example, that red meat increases the likelihood of developing cancer, journalists often ask researchers whether this means that individuals should eat less meat. This question implicitly also asks whether health is primarily a personal responsibility, or whether the food industry and the government have an obligation to encourage health. Similarly, when climate change came to the fore as one of the main challenges of our time at the beginning of the twenty-first century, public debates focused on the extent to which blame should be placed on individuals with a large carbon footprint.

In part, these questions concerning responsibility for risk are engendered by different notions of causality. For example, do individuals cause their own health problems by poor lifestyle choices and preferences? Alternatively, is it the fast-food chains that cause, for example, obesity by marketing unhealthy food and being so widely and easily available? Causality can be very complex and can be affected by contributing factors including natural laws, institutions, the economy and culture, depending on the level at which the question is framed. Furthermore, opinions often differ as to whether a particular factor or individual has had a causal role in a given chain of events. One additional reason for such differing views on responsibility distributions is that there are two moral values at stake: fairness and efficacy. These values correspond to two different norms: namely, that responsibility should be distributed fairly and that it should be distributed effectively.

Two values

First, it could be – and frequently is – argued that responsibility ascriptions should be fair. For example, if the person being held responsible had nothing to do with the incident for which she is held responsible, this will not be perceived to be fair. Oftentimes, the perception of unfairness stems from the fact that it is less clear who did and who did not cause a series of events. We might also not consider it fair to ascribe responsibility to someone who cannot do anything about that particular risk. Thus, evidence of some kind of causal link is a reasonable requirement for fair responsibility ascriptions.

Second, the question as to who is responsible for something usually arises because there exists a problem in need of a solution. Something needs to be done, and, by ascribing responsibility to one or more agents, a division of labour is established, the aim of which is to facilitate a solution. There is, in a sense, a moral imperative that something ought to be done, and the distribution of responsibility distributes the "oughts" in a – hopefully – reasonably effective way.

Ideally, responsibility distributions are both fair and effective. However, the two values may conflict; and, whereas some ethical and political theories emphasise fairness, others find efficacy to be the superior value. The primary aim of this chapter is to achieve more clarity regarding the two values. The tension between the two values is illustrated in the domain of public health, which is an area where questions of individual responsibility are particularly salient.

There are two major philosophical perspectives on responsibility ascriptions, and they view the two values quite differently. First, there is the merit-based perspective that responsibility should be ascribed to someone who *deserves* to be held accountable because she intentionally and knowingly brought about her own problem. The idea then, is that risks are chosen by the individual, which arguably makes her responsible for the outcomes. Therefore, the argument goes, if you choose to smoke in spite of the risks, you deserve to be held responsible for developing lung cancer. This perspective is based on the notion that fairness is the overriding value in responsibility distribution.

Second, there is the consequentialist perspective, which asserts that effects are the only thing that matters when ascribing responsibility. A utilitarian version of this view would entail that responsibility should be ascribed in such a way that utility is maximised. The following two norms each focus on one of the values and are based on opposing philosophical perspectives:

A) Deontological (based on notions about duties and rights) norm: a responsibility ascription or distribution ought to be fair; that is, based on merit and that which is deserved.

B) Consequentialist (based on the notion of optimisation of consequences) norm: a responsibility ascription or distribution ought to be effective; that is, have the best possible consequences compared to other responsibility ascriptions or distributions.

Intuitively, we can see that, when responsibility is distributed, it is important to take both values into account, and, consequently, that both norms are reasonable (Nihlén Fahlquist, 2006). When holding someone responsible, there is often an implicit assumption that this is fair *and* will contribute to solving the problem at hand. It is natural and an important aspect of being a member of a moral community to hold oneself and others responsible for actions. Fairness could be, and frequently is, defined in different ways, but regardless of the exact definition of "fairness", the notion here is merely that whatever is defined as fair is more important than efficacy or consequences. We may not think about this when actions are ethically "neutral", but when they appear to be right or wrong we do. Hence, when individuals do things that appear to us as morally right or wrong, we praise or blame them. As Strawson describes it, we have expectations of others and of ourselves, and when we do not live up to these expectations, we react with emotions like remorse, feelings of responsibility, guilt and resentment (Strawson, 1962). These emotions are all associated with a notion of fairness and the right to be treated in a certain way. As we saw in the introduction, Fischer's revised version of Strawson's theory states that "agents are morally responsible if and only if they are *appropriate* recipients of reactive attitudes" (Fischer and Ravizza, 1993). Questions of fairness arise when we reflect on who an appropriate target of reactive attitudes in a specific situation or context is. Furthermore, fairness becomes relevant when some kind of distribution of responsibility will take place.

However, when analysing risk it is reasonable to think that efficacy matters when distributing responsibility. If we consider criminal liability as representative of the social problems of risk, it is reasonable to assume that punishments deter criminals and therefore hopefully restore societal function. In the context of societal decision-making concerning risk problems, efficacy is even more important since we aim to achieve a certain result, i.e. for example a risk reduction.

Responsibility distributions can be more or less formal. The kind of responsibility distribution that we do on an everyday basis when we form reactive attitudes to fellow human beings is informal. In the context of societal decision-making, responsibility is distributed in a more formal way.

Public health: inherent tension between fairness and efficacy

Depending on which value is prioritised, different conclusions may follow. Public health is a good illustration of the potential tension, or even conflict, between the values of efficacy and fairness. Controversies exist about the main purpose of public health and about why people's lives ought to be prolonged, or their health promoted, through public health interventions. Arguments for health-promoting interventions in people's lives can be divided into two categories: those that emphasise the well-being of the individual citizen, and those that focus on the social costs of bad health (Vallgårda, 2001). The former arises from a deontological paradigm in which individual rights and fairness are emphasised. Notions of equal rights to health belong to this group of arguments. In contrast, arguments concerning the social and financial costs of poor health are consequentialist and focus attention at the societal level and on collective risk reduction and utility, understood as quantifiable health (Nihlén Fahlquist, 2006). Public health is a modern invention. Previously, healthcare was restricted to the context in which patients saw doctors in a one-on-one setting. The introduction of public health implies an emphasis on the value of collective health improvements, which may not always correspond to individual health improvements or actualisation of individual rights. It is thus essentially a utilitarian project, meaning that the goal is a general maximisation of utility in terms of improving the health of the population (Horner 2000).

Although not everyone would agree that public health is "utilitarian" or that health is a "utility", the goal of public health is necessarily the improvement of the overall health of the population rather than the individual. Therefore, when assigning responsibility for public health problems, it is reasonable to require that responsibility ascriptions be effective, and that they should contribute to a positive change, such as a reduction of mortality or risk. We will now look at the two opposing theories that emphasise fairness and efficacy respectively in responsibility distributions in the context of public health.

Libertarianism and rights – fairness as only relevant value

Libertarian views on the importance of respecting individual liberty will serve as an example of a merit-based perspective on how to ascribe responsibility in the area of public health.

According to libertarian thinking, individuals are free moral agents and, as such, responsible for their own actions. Hence, they should be allowed to decide for themselves whether or not to engage in unhealthy behaviour. At the same time, the government in turn has no responsibility to pay for their healthcare. Individual liberty is morally and politically central to libertarianism, and all libertarians reject all forms of coercive public health, such as high taxes on certain food and beverages. The only state that is legitimate, according to Nozick, is the minimal state, because a more extensive state "will violate persons' rights not to be forced to do certain things" and would therefore be unjustified (Nozick 1974).

According to this view, a government engaging in public health activities is not morally legitimate and does not respect the rights of individuals. It would not be fair to force individuals to participate in disease-preventing or health-promoting activities. The only fair distribution of responsibility for health is that individuals are completely responsible for their own health, and not responsible for anyone else's health. Requiring people to pay taxes in order to provide a state with the funds necessary to engage in public health activities in order to improve the general health would infringe on individual rights, because it would not honour the Kantian idea (which Nozick's theory endorses) that human beings should be treated as ends-in-themselves. Is this the correct conclusion? Perhaps it is not the whole truth, because although most libertarians would strongly disagree with coercive measures such as taxes on allegedly unhealthy food, some libertarians do support restricted information campaigns to raise awareness. The concept of "nudging" has become very popular during recent years. Originally, Thaler and Sunstein introduced the concept to illustrate how government action can gently nudge people in the right direction, affecting their choices without substantially limiting their freedom. Normally, libertarianism – as proposed by, for example, Nozick – is taken to exclude all infringements of individual rights and liberties. This argument is often taken to prohibit any interventions into freedom of choice. However, Thaler and Sunstein challenge this idea by showing that even libertarians must commit to a certain measure of paternalism because someone has to present the options in some way or another. Against this background, the options might as well be presented and framed in a way that affects people's choices in a way that promotes e.g. good health rather than poor health. The most basic example they provide is the way a cafeteria in a workplace must choose whether to put the fruits and vegetables first or the sweets in front. If fruits and vegetables are placed in a way that makes them visible and attractive to hungry people when standing in line, more people are likely to choose them. Assuming that greater availability of fruit and vegetables will lead to more people eating more fruits and vegetables and fewer sweets, people will become healthier without having their freedom of choice severely diminished. According to Thaler and Sunstein, a "choice architect" is anyone who influences the choices we make as human beings in modern society. Since choice architects cannot avoid presenting options in one way or another, they might as well do it in a way that could lead to improvements in people's decisions. In this vein, even libertarians would agree that the government is allowed to affect people – to nudge them – in a non-intrusive way. Even libertarians would, the argument continues, agree to a limited form of paternalism (Thaler and Sunstein, 2008).

However, in this context we are merely concerned with the strict libertarian view, which excludes public health interventions. The conclusion of a strictly libertarian way of reasoning about public health is that, from a moral point of view, individuals are responsible for their actions, and whether they should smoke or not smoke is their decision, regardless of the public health problems that may result. Even if it would prove more effective to ascribe responsibility for health to the government, that should not be done, according to this line of reasoning. Efficacy, in this view, is irrelevant. The only value that matters is fairness, and the only fair way to distribute responsibility for health is to say that individuals are completely responsible for their own health.

This is a simplified version of libertarianism, presented in order to clarify the different norms and perspectives. Things, and theories, are always more complicated in reality. One challenge to libertarians is where to draw the line between individual harm and harm to others, since individual liberty is only to be restricted when causing harm to others (cf. Mill, 1865; Feinberg, 1984). The notion of the individual who is solely responsible for herself and who, as long as she does not infringe on the rights of others, can do whatever she wishes, has been severely challenged during recent decades. If we consider infectious diseases or vaccination, it becomes clear that an individual's choices concerning her own life and health necessarily affects others when living in a society. Human beings are social animals, and very few of our actions do not affect others in any way. The idea that individuals are not allowed to infringe on the rights of other individuals makes behaviour like smoking problematic. On the one hand, individuals have the right to smoke because they are allowed to do what they want to do regardless of whether they harm their own health. On the other hand, they are only allowed to do so if they do not infringe on the rights of other individuals. They do not have the right to impair other people's health. Since smoking is a health problem for non-smokers, who suffer from passive smoking, there is a conflict between the rights of active smokers and the rights of unwilling passive smokers. The same problem faces the libertarian who wants to argue for the right to drive as fast as one wants, which obviously affects a lot of people and can be said to infringe on their rights. The libertarian view serves as an example. However, one need not adopt a libertarian outlook in order to argue that individual rights and fairness are paramount in the context of responsibility distributions.

Utilitarianism – efficacy as only relevant value

In contemporary society we are usually not only interested in fairness when discussing risk and responsibility. In modern democratic states, loosely defined, we also ascribe responsibility in order to achieve progress and to solve problems collectively. A strict utilitarian would make health a personal responsibility if that were the most effective way to solve health problems. If that would not be the most effective way, she would not.

A common method used in public health policy is cost-benefit or risk-benefit analysis, a quantitative approach that defines risk as probability of adverse outcomes multiplied by unwanted effect. The goal is to weigh the options in terms of costs or risks on the one hand and benefits on the other. The option that leads to the most benefits and the least cost or risks is then considered preferable. Policy-makers use this method to make decisions about what should be done, and which risks and policies should be promoted. Cost-benefit analysis is a utilitarian method. If applied consistently, fairness would not be considered unless it affected the total utility.

As mentioned previously, public health is essentially a utilitarian project in that its rationale is to maximise the health of the people. However, in practice, public health authorities cannot solely be concerned with the utilitarian aim of their activities, but must respect human rights as well, and human rights are based on deontological notions. Rights-based perspectives and related notions of social justice have become highly salient in public health debate during recent years (e.g. Daniels 1985; Powers and Faden, 2006).

Efficacy and responsibility for health

Against this background, one of the fundamental questions in the context of public health is how to reconcile the rights of the individual and the interests of society (Horner, 2000). This is another way of expressing the tension between fairness and efficacy. This is not an easy task and can be ethically controversial.

One example of this is, as mentioned above, infectious diseases, where it is sometimes necessary to override the autonomy of individuals to reduce the risk of the disease spreading to the rest of the community (Horner, 2000). For example, consider vaccination. If people stop having their children vaccinated against measles, the disease may come back and cause substantial harm. Do individuals have the right to refrain from having their children vaccinated? Or should the value of efficacy lead states to legislate on mandatory vaccination? A third example is taken from the context of cancer prevention. A long-standing research question seeks to determine the extent to which cancer is caused by genetic factors as opposed to lifestyle or environmental factors. Recent studies estimate that a large proportion of the most prevalent cancer types are primarily caused by environmental factors, indicating that cancer prevalence could be modified through lifestyle changes and vaccinations (Wu *et al.*, 2016).

This clearly raises the question of personal responsibility for health. If we take the example of smoking, it is generally well known that it causes serious health impairments and, in many cases, leads to death. In spite of this, some people continue to smoke, resulting in negative health effects and expensive healthcare, in which case taxpayers' money (in liberal European welfare states) will be used to pay for health expenses that could have been avoided if all smokers gave up smoking. Should we take social action to prevent smoking, thereby protecting these people's health, when they clearly do not understand or value their own, or society's, best interests? If so, there are several, more or less intrusive, ways of doing that. People's behaviour can be influenced through information campaigns or access to stop-smoking programmes. Additionally, the government can legislate in order to increase taxes or to restrict smoking in public spaces, thereby delimiting the choices of smokers. Alternatively, should we refrain from such actions, adopting the attitude that smoking is a problem for which smokers are themselves responsible? What would then be the most effective way to deal with the problem? Meanwhile, some people get lung cancer without having smoked, which highlights the need to take the fairness perspective into account and raises the question of how to avoid stigmatising lung cancer patients.

From the efficacy perspective, there is clearly a balance to be struck here. If we assign too much responsibility to the individual in cases like obesity or smoking, we may not get the results we want. People could be stigmatised, avoid the message, or resent the government's attempts to communicate risks. This approach does not appear to be effective. On the other hand, assigning all responsibility to the government could also be counterproductive since this may discourage individuals from doing their best to avoid health-impairing behaviour. The effective assignment of responsibility in order to get the results we want, whether it is a reduction in injuries and other health impairments or a reduction in fatalities, is ultimately an empirical question. This contrasts with the value of fairness, where empirical investigations cannot answer the question of how responsibility should be distributed.

To see how responsibility ascriptions can be more or less effective, let us consider another hypothetical example from healthcare. Imagine a nurse who has been instructed to administer 0.05 mg of the medical substance X to a patient. He accidentally gives the patient 5 mg of X, and, as a result, the patient's condition worsens. Who is morally responsible? At first, we may say that the nurse is responsible. However, if we find out that he is a conscientious and hardworking professional who is exhausted because of stress and long working hours, and that the containers filled with 0.05 mg and those filled with 5 mg look exactly the same, we may decide to assign some responsibility to his managers and to the company that produced the containers. If he is a competent nurse, assigning all responsibility for this accident to him and dismissing him would not solve the problem. Other nurses may make the same mistake in the future. It could be argued that it is more effective to assign some responsibility to the managers and drug suppliers as well, and perhaps to consider the nurse to be only partly responsible.

As we can see in the case of the nurse, it is important to distinguish between triggering causes and systematic or structural causes. The nurse might have triggered the harm, but there may be structural or underlying causal factors having to do with working conditions, for example, which should be addressed as well in order to avoid similar incidents in the future. Experiences from aviation safety may serve as an additional example of how responsibility ascriptions can be more or less effective. Traditionally, airplane incidents or accidents have been blamed on the pilot – in other words, on the "human factor" or "human error". Human factor researchers have been questioning this approach to assigning responsibility since the 1990s because it is allegedly not effective, since pilots knew that if they reported mistakes and errors they had made, they might be blamed. Consequently, such incidents were not reported. This could, of course, potentially lead to more accidents. To avoid this unintended effect and to improve safety, pilots in some countries were permitted to report errors anonymously so that information could be gathered without blaming pilots. Focusing on the person who was most responsible for causing the accident – the pilot in this case – often makes us overlook the underlying causes of the incident (Beaty, 1995; Dekker, 2002). This illustrates that the way in which we assign responsibility can determine the way in which complex problems are addressed. There clearly are more or less effective ways to assign responsibility. Emphasising the value of efficacy necessarily implies a consequentialist approach to responsibility.

Fairness and efficacy in policy-making concerning risk

As we have seen, there is an inherent tension in responsibility debates between fairness and efficacy. First, we want to assign responsibility to the agent who deserves to be held responsible so that the distribution of responsibility is justified and morally right. Second, there is usually a problem that needs to be solved by assigning responsibility to someone who can actually make a difference. Some of us tend to emphasise the moral content of responsibility ascriptions and believe that we cannot legitimately ascribe responsibility to someone who is not blameworthy. Others point to the importance of a problem-solving approach to responsibility ascriptions.

The application of cost-benefit analysis to the question of how to distribute responsibility leads to the following approach to addressing the question of responsibility distributions: if the benefits of individual responsibility for health are greater (and/or the costs lower) than the benefits of ascribing responsibility to the government and/or industry, then this is the way we ought to distribute responsibility. Alternatively, if the benefits of ascribing responsibility to the government and/or industry are greater than ascribing responsibility to individuals, we would do that instead.

Arguably, most of us are neither pure consequentialists nor strict deontologists but recognise the importance of both values relevant to responsibility ascriptions. In society, the most reasonable way to ascribe responsibility may be to take both values into consideration; in other words, to aim for a distribution of responsibility that balances the values of fairness and efficacy.

If it is true, or highly likely, that between 70 and 90 per cent of cancer cases are caused by lifestyle, does that mean that individuals who are diagnosed with cancer, or perhaps 70 to 90 per cent of those people, are responsible for their own disease? Stating that it does mean that could, in turn, mean different things. It could mean that they caused their own cancer, that they are to blame for it, and/or that individuals are the ones who are responsible for avoiding cancer-promoting behaviour. A research finding, such as the one mentioned above indicating that the major part of cancer cases are due to lifestyle, could also be discussed from the perspective of the two values. First, would it be effective to hold individuals responsible for choosing behaviour that is likely to lead to cancer? Stressing individual responsibility could lead to a society in which people take all necessary measures to avoid cancer-promoting behaviour, including quitting smoking and exercising. If this approach were effective, would it also be fair? Do people who choose to smoke deserve to be held responsible for developing lung cancer? That is a very different question. We end up with the following separable questions:

1 Is it fair to hold smokers responsible for having lung cancer?
2 Is it effective to hold smokers responsible for having lung cancer?

Perhaps the best responsibility distribution would simultaneously avoid blaming and stigmatising those individuals who already have cancer and would encourage people who currently do not have cancer to refrain from cancer-promoting behaviour. Whether this is possible in practice remains to be seen.

The tension between efficacy and fairness approaches to responsibility ascriptions is not likely to be resolved but being aware of it may help clarify areas of agreement and disagreement. Having acknowledged this distinction between being responsible and holding someone responsible, the aim in policy-making should arguably be to strike a balance that is both perceived as fair by most citizens and is at least reasonably effective.

This requires that both perspectives be taken into account in public health policy-making. A reasonable requirement in policy-making is to expect the two approaches to responsibility ascriptions to be acknowledged and that a decision to let fairness outweigh efficacy or vice versa should be explicit and accompanied by evidence and good ethical arguments. One further implication is that the dominant approach of using cost-benefit analysis as a tool to decide which risks to accept and which to reduce needs more justification than what is usually provided. When cost-benefit analysis is used, efficacy is the crucial norm. It is not always clear if or why this is the preferred way to go, from an ethical perspective. It may sometimes be pragmatically and/or economically preferable, but this is not the same thing as saying it is ethically preferable.

Conclusion

Fairness and efficacy can be separated in theory, as can the individual and social aspects of complex issues such as public health, but the reality is obviously more complex. Although the individual may, to some extent, choose to eat an unhealthy diet, to smoke, or to avoid exercise, measures taken by public authorities and by industry also have an effect on individual health. Similarly, an individual chooses to drive a car or to use public transport, but she acts within an infrastructural context that is the result of political decisions and societal risk management. Additionally, socio-economic factors play an important causal role in health differences (Anand, Fabienne and Sen, 2004). These factors are also important in terms of fairness in distributing responsibility for global concerns such as climate change (Nihlén Fahlquist, 2009). Thus, we are dealing with multi-faceted problems where the actions and omissions of individuals as well as of society, the government and industry, contribute both to negative and positive outcomes in terms of mortality and morbidity (Nihlén Fahlquist, 2006).

The combination of individual and social aspects of responsibility ascription creates a moral problem for policy-makers. If we take public health as our example, the central ethical dilemma is precisely this conflict between the rights of the individual on the one hand and society's responsibility for all its members on the other, because the former lends itself to an emphasis on fairness, while the latter lends itself to a utilitarian focus on efficacy. Another way of putting it is in terms of justice, in order to clarify how the two are interconnected. On the one hand, each individual arguably has a right to health. As asserted by Daniels, there is a right to healthcare because health impacts the extent to which the liberal goal of equal opportunities can be realised. If there are large health inequalities, people cannot be said to have equal opportunities. Therefore, universal healthcare is needed to maintain normal functioning (Daniels, 1985). On the other hand, there are not enough resources to provide all members of society with "perfect" health, which means that society must prioritise and spend money effectively and efficiently. If each individual has a right to health (fairness perspective), we cannot spend state money ineffectively (efficacy perspective) because that will lead to less money to ensure the rights of each individual. In this sense, fairness requires efficacy. Consequently, the relationship between the two values is very complex.

Because of this, it appears reasonable to ascribe responsibility to the individual, the government and industry, but the content, nature and implication of responsibility may differ depending on circumstances. Some participants in the debate on, for example, the taxation of unhealthy food, appear to assign all responsibility to individuals, whereas others seem to assign a large part of responsibility to public authorities or to the food industry. Therefore, it is important to acknowledge the complexity of the matter and to clearly separate the values at stake. In policy-making concerning risks, the aim should arguably be to acknowledge both values and to attempt to find a way to reconcile them.

Note

1 Some parts of this chapter are reprinted with permission from the following article: Nihlén Fahlquist, J. 2006. Responsibility ascriptions and public health problems. Who is responsible for obesity and lung cancer? *Journal of Public Health* 14(1): 15–19.

References

Anand, S., Fabienne, P. and Sen, A. (eds). 2004. *Public Health, Ethics, and Equity*. Oxford: Oxford University Press.

Beaty, D. 1995. *The Naked Pilot: The Human Factor in Aircraft Accidents*. Shrewsbury: Airlife.

Daniels, N. 1985. *Just Health Care*. Cambridge: Cambridge University Press.

Dekker, S. 2002. *The Field Guide to Human Error Investigations*. Aldershot: Ashgate.

Feinberg, J. 1984. *The Moral Limits of the Criminal Law* (Vol. 1 of *Harm to Others*). New York: Oxford University Press.

Fischer, J.M. and Ravizza, A. (eds). 1993. *Perspectives on Moral Responsibility*. Ithaca, NY: Cornell University Press.

Horner, J.S. 2000. The virtuous public health physician. *Journal of Public Health* 22: 48–53.

Mill, J.S. 1865. *On Liberty*. London: Longman.

Nihlén Fahlquist, J. 2006. Responsibility ascriptions and public health problems. Who is responsible for obesity and lung cancer? *Journal of Public Health* 14(1): 15–19.

Nihlén Fahlquist, J. 2009. Moral responsibility for environmental problems – individual or institutional? *Journal of Agricultural and Environmental Ethics* 22(2): 109–124.

Nozick, R. 1974. *Anarchy, State and Utopia*. Oxford: Blackwell.

Powers, M. and Faden, R. 2006. *Social Justice*. New York: Oxford University Press.

Strawson, P.F. 1962. Freedom and resentment. *Proceedings of the British Academy* 48: 1–25 (reprinted in J.M. Fischer and M. Ravizza (eds), *Perspectives on Moral Responsibility* (pp. 45–66). Ithaca, NY: Cornell University Press, 1993).

Thaler, R.H. and Sunstein, C.R. 2008. *Nudge: Improving Decisions about Health, Wealth, and Happiness*. New Haven, CT: Yale University Press.

Vallgårda, S. 2001. Governing people's lives. *European Journal of Public Health* 11: 386–392.

Wu, S., Powers, S., Zhu, W. and Hannun, Y.A. 2016. Substantial contribution of extrinsic risk factors to cancer development. *Nature* 529: 43–47.

2 Backward-looking and forward-looking responsibility

Introduction

In the previous chapter, assignment of responsibility was discussed from the perspective of fairness and efficacy, which are two central values in responsibility assignment and distribution. As we could see in the example of public health, one norm when ascribing responsibility is to make sure it is fair, and another norm is that it should be effective. Fairness can be defined in many different ways and can be seen as an essentially contested concept; that is, a concept where there will always be a certain degree of disagreement (Gallie, 1955). From a deontological perspective, fairness is the overriding value, while the consequentialist perspective focuses on efficacy, or how to solve a particular problem, such as increasing obesity. Thus, one way of evaluating responsibility assignment and distribution is to analyse to what extent they are fair and to what extent they are effective.

We will now explore a second important distinction in the analysis of responsibility assignment and distribution – namely, between two kinds of responsibility assignment, referred to as backward-looking and forward-looking responsibility. This distinction is very useful when analysing responsibility for technological, environmental and health-related risks. In order to see the relevance of this distinction, consider the debate that often follows major accidents, such as the Deepwater Horizon oil spill in 2010. When a catastrophic event such as this occurs, the public usually has – explicitly or implicitly – four important, but rarely clearly distinguished, concerns regarding responsibility. First, what and who caused it? Second, who is to blame? Third, how can it be prevented in the future? Fourth, whose task is it to do so? Whereas the first two questions are primarily backward-looking, the third and fourth are forward-looking. Sometimes, these questions are conflated, but it is important to keep them separate since it is not always reasonable to answer all of the questions by referring to one and the same agent.

In this chapter, we turn our attention to the distinction that arises from these questions of backward-looking and forward-looking responsibility.

In addition to asking ourselves how to balance the values of fairness and efficacy when distributing responsibility for risks, we should also decide whether, or to what extent, we are distributing backward-looking or forward-looking responsibility.

Causing the mess and picking up the pieces

Sometimes, the agent who caused something – and who is potentially blameworthy – is also the one who should do something about it. For example, if Anna throws a vase at Beth because she is mad at her, our holding Anna responsible (if we assume we have all the information needed for a moral assessment) may address both issues. We would perhaps blame her *and* expect her to apologise and clean up the mess. However, there could be other agents who are also responsible to some extent or in some way. If it turns out that Anna has a psychosis, which is why she threw the vase, and that Charles has the power to help Anna get the proper care, he may have an additional responsibility in that situation to make sure Anna gets proper psychiatric care. He has this responsibility because, for instance, Anna knows and trusts him because he is her husband. Furthermore, imagine that he knows how to get proper care because he also a general practitioner. His responsibility became relevant as a consequence of the event that took place. Possibly, when Anna threw the vase at Beth, it became clear that something was wrong with Anna's mental health, since there was nothing external to cause her anger. However, Charles may not have had anything to do with Anna's throwing the vase at Beth. In this case, Charles may be considered to have an obligation or responsibility to make sure Anna gets the proper care regardless of whether Anna does or does not have backward-looking responsibility for what happened. This example highlights the complexity of the concept of responsibility and the relevance of contextual considerations. It also illustrates the need to distinguish between backward-looking and forward-looking responsibility.

Taking another example from risk management, we could look at the context of aviation. Even though an individual pilot may have caused an accident, the company for which she works may have additional obligations requiring a change in working conditions or technological design in order to prevent pilots from repeating the same "human error". This example also illustrates the difference between the backward-looking and forward-looking dimensions of responsibility.

Thus, the two kinds of responsibility could, but do not *necessarily*, coincide. For this reason, we need to maintain the distinction between backward-looking and forward-looking responsibility. This distinction is sometimes made in scholarly literature. Bovens, for example, makes a distinction between passive responsibility, which concerns being *held* accountable, and active responsibility, which concerns *taking* responsibility and is based on virtue ethics. In the former case, the question is "What did you do?" In the latter case, the question is "What is to be done?" (see Bovens, 1998). Bovens argues that responsibility as task constitutes a third kind of responsibility, one that forms the "connective element" between the two others (ibid.). There are various ways of distinguishing different notions of responsibility, and I would suggest that, whereas accountability is backward-looking, responsibility as virtue and responsibility as task are the two main categories of forward-looking responsibility. We will explore responsibility as virtue in Part II. For now, it suffices to say that there is a fundamental conceptual distinction to be made between responsibility for what happened and responsibility for what is to come, and this distinction could have normative implications.

Causal responsibility and blame

Discussions concerning responsibility are often heated partly because of different normative perspectives that influence our views of who is responsible. For example, whereas liberalism and libertarianism locate a great share of responsibility with individuals, Marxists place a greater share of responsibility with collective actors. This difference of perspectives is inherent in the decision to base the analysis of a given situation on individual or structural causes, because each perspective entails a different approach to responsibility. Partly, this relates to different perspectives on fairness and efficacy, as we saw in the previous chapter. In addition, I would say that our different perspectives regarding responsibility depend on conceptual ambiguity. For example, when stating that "X is responsible for Y", are we looking for the agent who caused Y, who is to blame for Y, or who could and/or should solve the problem? Due to this ambiguity concerning what we actually mean, it is useful to begin the analysis of backward-looking and forward-looking responsibility assignment by distinguishing between causal responsibility, blame responsibility and forward-looking responsibility.

Sometimes, causation is conflated with blameworthiness. For example, as Smiley argues, we do two things when we ascribe moral responsibility to an agent for a particular event. First, we assert something about the causal background of, for example, the accident. Second, we blame the agent if said agent is thought to have acted voluntarily and knowingly (Smiley, 1992). Imagine that there has been a car accident involving a drunk driver who severely injured a child. In such a case, we would probably say that the child was injured *as a result* of the driver's intoxication and then apply a set of social and moral norms regarding the responsibilities of drivers This driver would probably be blamed, because she decided, more or less consciously, to risk her own and others' lives by driving under the influence, an action that is considered morally wrong in most societies. However, "the cause" could also be more complex. If the drunk driver is an alcoholic who has not been given the right treatment, this could also be seen as part of the cause. Furthermore, if there is technology, such as the alcohol interlock that could eradicate drunk driving, a government's failure to legislate may also be seen as part of the cause.

Conditions of backward-looking responsibility

Usually, when backward-looking responsibility is ascribed or distributed, a set of conditions is, explicitly or implicitly, applied. These conditions go back to Aristotle, who described what features of a situation were important when blaming and praising someone. Without using the term "responsibility", he focused on the question of what aspects of an act made it voluntary, and, therefore, potentially blameworthy or praiseworthy. He argued that if an act is to be considered voluntary and therefore blameworthy, it must have been done freely and knowingly. In other words, it is not right to blame X for doing Y unless she did it voluntarily; that is, freely and knowingly. If we consider climate change, it is reasonable to think that human beings were not voluntarily causing it when we did not know that climate change was anthropogenically caused. However, during the last few decades we have known that we, as human beings, cause climate change by carbon dioxide emissions, which means that we are possibly blameworthy today when contributing to climate change or failing to mitigate it. However, having said that human beings are potentially blameworthy for causing climate change, we have yet to say how that responsibility should be distributed between different actors, such as individuals, governments and industries.

To this day, it is common in everyday situations, as well as in philosophical literature, to ascribe backward-looking responsibility based on a number of conditions. The following is a list of conditions that are often seen as prerequisites for holding someone morally responsible in the backward-looking sense (Van de Poel *et al.*, 2012; Nihlén Fahlquist, 2017b):

- Moral agency/capacity.
- Causality.
- Knowledge.
- Freedom.
- Wrongdoing.

The essential notion then is that when stating that "X is responsible for Y", this is short for stating that "X is a moral agent, who caused Y, which involved doing wrong, and she did it freely and had the relevant knowledge". Let us go through these conditions in turn.

Moral agency/capacity

The first condition requires that the potentially responsible agent is a moral agent who has a certain capacity for rational thinking and moral deliberation. The boundaries and conditions of moral agency are not set in stone but have varied throughout history and by context. For example, it has recently been suggested that children should be considered moral agents (cf. Mullin, 2007) and they are, in some contexts. For example, in Belgium, children have the right to choose euthanasia. However, in many contexts, children are not considered autonomous moral agents. In most countries, young children cannot be incarcerated because of their cognitive limitations. The question of children as moral agents in the context of risk and responsibility is explored in Chapter 6. Similarly, severely cognitively impaired people are usually exempted from responsibility, although it is often not clear where the line is to be drawn. People suffering from dementia are a good case in point when being faced with the choice of medical treatments. Are they still to be considered autonomous? Are they moral agents in the same sense as they were before they developed dementia? If not, when did they stop being moral agents? These questions are complex and illustrate the difficulties in drawing the line between moral agents and subjects who may be deserving of respect as persons, but who are not to be considered responsible for their actions. In the same vein, there is currently a debate about the potential agency of robots (Johansson, 2010). Being a moral agent means being a member of the moral community and is a requirement for blameworthiness and accountability. However, it is not necessary to be a moral agent to be considered morally important, i.e. for example, as someone who others ought to treat with respect and avoid harming. Animals, for example, could be seen as having moral status requiring human beings to treat them well without being moral agents. Who is considered to have moral agency and moral status varies over time.

Causality

Even though "causality" is partly a scientific concept, it is also social and normative in some contexts. Although a risk may have materialised due to many different underlying factors that can be described as "causes", when we talk about these factors, one (or more) is usually singled out and described as "*the* cause" (Smiley, 1992). Which of the causally relevant circumstances we single out as the most important is determined by our norms, goals and prior assumptions; thus, it is to some extent dependent on the individual and the social context. As in the example with the traffic accident mentioned above, one norm could be that drunk driving is morally wrong, and when an accident involves a drunk driver the fact that she was intoxicated is likely to be seen as a major cause of the accident if there are social norms condemning drunk driving. It is also imaginable that a particular culture is harsher towards either female or male drivers who make errors, which could contribute to the extent to which they are blamed for those errors. There are often many necessary contributing conditions, and perhaps none of them would be sufficient on its own in any given case (Downie, 1971). This means that an event can most probably be described as the result of different causes depending on who is asked to explain it (Feinberg, 1970). For example, climate change may be described as the consequence of emissions of carbon dioxide, or of the institutional and political failure to deal with the problem, or of overconsumption and our current way of living, or something else. Similarly, obesity could be described as a consequence of overeating, genetic makeup, lack of exercise, car dependency, or availability and affordability of unhealthy food. It could be described as a consequence of individual risk-taking or societal inability to manage lifestyle-related risks.

Thus, social, moral and political considerations often influence how we refer to causation. Statements of causal responsibility and blameworthiness are conflated into one "causal discovery" (Smiley, 1992). The concept of responsibility is used both to claim a factual connection between an agent and some state of affairs, and to assign individual accountability for bringing it about, which is normative (Dworkin, 1981). Thus, we seem to think that if agent A is causally responsible for harmful event E, then she is also blameworthy. For example, if a smoker is said to have caused her own lung cancer, it is commonly considered acceptable to blame her for it. When a risk materialises, as with the drunk driver causing a fatality, it is often argued that it is important to "*find*" the one who is responsible for it. It is not always clear from such a statement whether it is an attempt to find the causal links or to establish who is to blame for the event (Smiley, 1992). Alternatively, we could be looking for the person or institution who could prevent similar things from happening again. In many cases, in ascribing responsibility to an agent for some harm, we do not acknowledge our own role as blamers, but assume that we have found the *real* causes of harm (Smiley, 1992). In this way, the process becomes a matter of mixing facts and values. Thus, in asserting that agent X is responsible for undesirable event Y, we make the assumption that Y was caused by X's actions *and* that she is morally blameworthy for the harm she caused.

Thus, the causality condition is very complex. However, unless the agent is causally linked, in some way, to the action or consequence for which she might be held responsible, she is usually not held responsible. It has been argued that consumers in industrialised countries contribute to structural processes causing global injustice and that they are, therefore, responsible for that injustice (Young, 2006). The argument is that, by buying clothes produced in a high-risk environment in a poor country, I contribute causally to a system where such high-risk environments continue to exist. The way individuals contribute to climate change is another example of a relatively loose causal connection. However, it is clear that some more or less tangible causal connection is required for moral responsibility, regardless of the complexity of the cause–effect relationship.

Voluntariness: freedom and knowledge

As we saw above, the conditions of knowledge and freedom have been seen as prerequisites for assigning moral responsibility and blame since Aristotle, as he considered the two related to voluntariness. The most common excuses that people offer when they think they are wrongfully held responsible are "I did not know" and "I was forced, I had no choice".

Freedom has proven to be a very important factor in people's acceptance of risk. Psychologists have shown that if an individual feels that she chose a certain risky activity, she may be much more willing to accept a negative outcome than if she feels that someone else exposed her to that risky activity (Slovic, 2000). Philosophers have discussed the normative importance of voluntariness, arguing that it matters ethically whether someone chose a certain risk or was exposed to it involuntarily (Roeser, 2006). Whereas choosing to smoke would perhaps be an acceptable risk, being exposed to passive smoking would not be acceptable. Some cases are less clearly separable. Does the alcoholic choose to drink alcohol, and is she therefore responsible for her own diseases which may come as a consequence? How free was she when choosing the risks associated with high consumption of alcoholic beverages? The line between what is chosen freely and what counts as being exposed to risk by someone else is not always clearly discernible.

Knowledge is of course highly important in the context of risk. The very definition of risk involves the notion that we do not have perfect knowledge. There is something unwanted that may or may not happen. When cigarettes were invented, we did not know the extent of potential negative effects for human health. Science has now shown the strong correlation between smoking and lung cancer. Smokers diagnosed with lung cancer today would have a hard time convincing people that they did not know that smoking may cause lung cancer. Lack of knowledge would not count as an excuse, and so they might be held responsible for their own lung cancer. At the same time, the fact that we do have this knowledge today may also make us hold tobacco companies and governments responsible for their lack of action in this area.

Wrongdoing

The last condition requires that there has been some wrongdoing. What counts as wrongdoing differs depending on ethical theory and contextual considerations. However, we do not hold others responsible in the sense of blameworthiness unless there appears to have been some kind of wrongdoing (Van de Poel *et al.*, 2012; Nihlén Fahlquist, 2017a). In the context of risk, let us illustrate by looking at climate change and sustainability. In the 1950s, driving a car or throwing cans in the trash were not considered to be acts of wrongdoing. In the twenty-first century, with the knowledge we have concerning the environment, it is considered to be wrongdoing in certain sections of certain societies. Similarly, it was not considered wrongdoing for a white man to hit a black slave in the Southern United States in the eighteenth century, but it is today. What is considered wrongdoing varies with ethical, cultural and political perspective, and also with time. This does not necessarily imply a relativistic view of ethics (e.g. (Roeser, 2010).

These conditions largely relate to the fairness value and to backward-looking responsibility. In other words, if the conditions are fulfilled, it is usually considered fair to ascribe backward-looking responsibility to the agent. As an example, we can debate whether the alcoholic may be held responsible for developing liver cancer by referring to the condition of freedom. Similarly, when we debate whether individuals should be held responsible for choosing products that have been transported across the world, thereby generating climate-damaging CO_2, is the knowledge condition fulfilled? Can we be said to know how all the products we consume contribute to climate change, and how far does our obligation to know extend? The conditions can be distinguished and explicated in theory, and we do use them in practice, but often implicitly, and not always with a clear separation.

Forward-looking responsibility ascriptions

Thus far we have been concerned with undesirable events that have already happened and risks that have materialised, such as car accidents or oil rig explosions. However, some responsibility assignment refers to future events, either to bringing them about or preventing them from occurring. An analysis of responsibility assignment that is concerned with risks and societal problems would not be complete without a notion of this kind of responsibility. This is particularly important in the context of risk and uncertainty since they, by definition, concern the future; that is, what *may* or *may not* happen.

The fact that some seem to focus on backward-looking responsibility, whereas others are more concerned with forward-looking responsibility, may have something to do with the difference between the two opposing philosophical perspectives of moral responsibility, introduced in the previous chapter. The first is the deontological perspective, which states that moral responsibility should be ascribed to someone only if she has done something blameworthy or praiseworthy. The second is the consequentialist perspective, which holds that responsibility should only be ascribed to someone if doing so will result in the desired outcome, such as improved behaviour. According to this line of reasoning, if holding an individual responsible for something does not result in a favourable outcome, we should refrain from ascribing responsibility to her. As we recall from the previous chapter, whereas consequentialists primarily consider the value of efficacy, deontologists emphasise the value of fairness.

Capacity

What does it mean to say that someone is responsible for something in the forward-looking sense? In the famous children's book (first published in 1947) about Pippi Longstocking, who is described as the strongest girl in the world, the main character gives money to thieves and states that when you are very strong, you must be very kind (Lindgren, 1978). This statement captures an intuition that most of us share, although it is not always made explicit.

In relation to climate change, Garvey addresses the intuition that rich countries ought to do more than poor countries to combat climate change. He briefly suggests that, just like "ought implies can", sometimes, "can implies ought" (Garvey, 2008). In my view, there is something highly reasonable in his brief statement. Sometimes, a certain principle of responsibility is based on both backward-looking and forward-looking aspects. For example, the so called "Common but Differentiated Responsibilities (CDR)" principle established by the United Nations Framework Convention on Climate Change in 1992 states that rich countries should bear a greater proportion of responsibility for climate change, supported by two arguments. First, rich countries have historically contributed more to climate change. Second, rich countries have a greater capacity primarily in terms of power and resources to solve these problems (Nihlén Fahlquist, 2009). Whereas the first justification emphasises the backward-looking concept of responsibility, the second is based on a forward-looking notion of responsibility. The latter is a reasonable idea and, I would argue, should be applied to individuals as well as to nations. If an individual is in a good position to do something to contribute to the solution of environmental problems, she has a responsibility to do so (Nihlén Fahlquist, 2009). If an individual is less capable of doing so, she is not responsible to the same extent. I will explain this idea in further detail in Chapter 3.

In addition, consider the idea of Corporate Social Responsibility (CSR); that is, the notion that private companies have a responsibility (which goes beyond their legal responsibility) to contribute to social causes. If so, one may ask if they should do it because it is good for business, or if they should do it regardless of the effect on business, or if the reason for doing it matters at all, as long as the consequences are good. A third example relates to the notion that rich countries have a responsibility to help poor countries in various ways: with money, education and technology transfer. In developed, democratic and well-ordered nations, it is a shared notion that the government should give aid to less developed countries. There is often disagreement about the size of aid and the exact content (money or self-help aid), but that there ought to be some kind of aid to less fortunate peoples is now a well-entrenched notion in developed nations (Nihlén Fahlquist, 2009).

Thus, there is a widely shared intuition in contemporary societies that if an agent, whether individual or collective, has the capacity, power and resources to contribute to solving a social problem, they have a responsibility to do so; in other words, that power and capacity entails responsibility. This kind of responsibility is forward-looking and fairly open-ended (Goodin, 1986). Consequently, it leaves open exactly what the agent ought to do and involves a certain degree of flexibility and permission to improvise.

Let us examine further the distinction between backward-looking and forward-looking responsibility by, again, looking at the context of climate change. I have argued that this forward-looking, open-ended responsibility that is based on capacity undergirds the way in which individuals as consumers and citizens are responsible for environmental problems (Nihlén Fahlquist, 2009). We will return to this context in the following chapter. Suffice it to say at the intuitive level for now that individuals arguably should do what they can in the context of their particular situation and capacity. The more resources, power and capacity an agent has, the better her ability to contribute to solving the problem. In such a context, it is fair to ascribe forward-looking responsibility to her. In this view, the causal links are less important than they are for assignment of backward-looking responsibility. Just as Bill Gates did not cause the social problems he contributes to solving by giving money to charity, the agents who are best placed to contribute to solving environmental problems are possibly not the ones who contributed the most causally to these problems. Their ability to do more can be due to several factors, most likely involving financial resources but also education, information and leadership skills in terms of a talent for influencing others. Whether wealthy states caused the poverty of poor states is controversial and up for debate. However, even if we think that poor states caused their own poverty, it is reasonable to think that wealthy states ought to help them, due to their capacity. In terms of risk, even if individuals may be said to, at least partly, cause their own alcohol-related diseases, the government may be said to have a forward-looking responsibility to prevent and treat such diseases and alleviate the harm experienced by those individuals.

As will be discussed, there are primarily two different notions of forward-looking responsibility that are relevant in the context of risk. First, there is a consequentialist notion that is related to task responsibility; that is, the responsibility to achieve a certain state of affairs. Another way of describing this notion of responsibility is to say that it provides a way to distribute the "oughts". It is one thing to decide what ought to be done and quite another to decide who should do it. Second, there is the notion of responsibility as a virtue; that is, the virtue of becoming a responsible person and actively taking responsibility for the future. We will explore responsibility as a virtue in the coming chapters. For now, I focus on responsibility as a task to achieve a certain result or state of affairs. This is the most relevant concept in the context of responsibility assignment and distribution because responsibility as virtue cannot be ascribed to someone else but is a development or process that happens within the agent.

Responsibility distribution as task distribution

Assignment of forward-looking responsibility as a task is very common. We all have tasks in our everyday life and in our professional lives as teachers, physicians, lawyers, chefs and firefighters. Furthermore, we are friends, sons and daughters, as well as parents. Forward-looking task responsibility is directed towards the future and is ascribed to someone because of the role they have in a family or society. In the policy-making context, it is usually about some sort of goal-setting and goal-achievement. In the context of risk and risk society, it could, for example, be linked to goals concerning risk reduction and risk management. It can be referred to as "reducing the risks related to climate change" or "managing the risks associated with infectious diseases", "minimizing the risk of nuclear accidents", or "decreasing mortality caused by malaria".

The intended goal is often open and general when ascribing forward-looking task responsibility, such as "a safe road traffic environment", which is another way of stating that risks should be reduced. However, it can include certain specific duties. Taking a goal like "achieving ecological sustainability" as an example, this is a more demanding goal than "reducing the risks relating to environmental pollution". Whereas duties may be a part of forward-looking responsibility, they are not identical. The difference between duties and forward-looking responsibility has been described as follows: duties refer to specific concerns that could be explicitly stated in advance, while forward-looking responsibility includes the responsibility for coping with surprises and uncertainty. In this way, it includes the possibility of changing the rules (Richardson, 1999). In addition, the difference between duties and responsibilities can also be described as follows: "Duties dictate actions. Responsibilities dictate results" (Goodin, 1986, p. 50). Duties require that the agent perform certain actions, while responsibilities "allow agents to choose between alternative actions having the same overall consequences" (Goodin, 1986, p. 53). When applied to risk, goals such as "creating ecological sustainability" allow discretion and improvisation. In contrast, "[r]educing the number of fatalities caused by road traffic in Sweden to five hundred per year" is a more specific and duty-oriented goal.

What about the role of causation in forward-looking responsibility assignment? To influence the future – to take action – requires the assumption that the agent has it *in her power* to see to it that the expected result is achieved. This power can be compared to the causation required for backward-looking responsibility, as it is what ideally triggers the desired state of affairs, for example the achievement of a goal.

Forward-looking responsibility, just like backward-looking responsibility, appears at least potentially to involve an element of blame. If A does not do what is expected of her, despite the fact that responsibility has been ascribed to her and that she has the power to act, she may reasonably be blamed. It appears that the main distinction to be concerned with is the one between backward-looking and forward-looking responsibility assignment, both of which include assumptions about causation and blame (Nihlén Fahlquist, 2006).

Although people often disagree, normatively, about who is responsible for what, some disagreements could simply be due to the ambiguity of the meaning of "being responsible". Against the background of the analysis provided so far, part of the confusion is the common tendency not to clearly separate the following questions:

- Who caused A?
- Who is to blame for A?
- Who should do something about A, or A1, A2 (where the latter are tasks or obligations arising from the context)?

Using these questions, we end up with the following two kinds of responsibility assignment:

> - Backward-looking responsibility assignment (consisting of backward-looking causal responsibility and blame).
> - Forward-looking responsibility assignment (consisting of forward-looking causal responsibility, potential backward-looking causal responsibility, and potential blame).
>
> (Nihlén Fahlquist, 2006)

In essence, this implies that one way of ascribing responsibility is to do it after the undesirable event has occurred, in which case we are concerned with the agent who caused it and who is possibly also blameworthy. Another way of ascribing responsibility is to establish, beforehand, who should achieve the desired result. In this case, what is maintained is that it is agent A's forward-looking causal responsibility to achieve result R. This means that A must have it in her power to actually influence events in such a way as to achieve R. In addition to asserting that A should contribute causally to achieve R, we are also establishing that if something goes wrong in such a way that R is not achieved despite the fact that A had been ascribed responsibility to achieve R and we have determined that it was in her power to achieve R, A will be held causally responsible in a backward-looking manner and possibly be blamed for the non-occurrence of R.

We will now look at the risks associated with road traffic, and how a Swedish policy brought about a new way of distributing responsibility that affected the balance between backward-looking and forward-looking responsibility.

Vision Zero – separating backward-looking and forward-looking responsibility[1]

In the month of September, 2001, more people died in car crashes in the United States than in terrorist attacks (Husak, 2004). Every year, more than 1 million people die globally in road accidents, and 20 to 50 million people are injured. Ninety per cent of road traffic fatalities occur in low- and middle-income countries, where it is a growing problem (Peden *et al.*, 2004; WHO, 2017). Globally, road traffic accidents are the main cause of death among those aged 15–29 years (WHO, 2018).

In 1997, the Swedish government decided to adopt a new traffic safety policy. It was called Vision Zero, referring to the visionary goal to have zero fatalities due to road traffic. The policy goal has been influential in many other countries. It was progressive in its insistence that it is not ethically justified to set a number of "acceptable" annual fatalities. This assertion should be compared to the cost/risk-benefit analysis usually used to guide public policy. Risk-benefit analysis would assume that we have to accept a certain number of fatalities because not doing so would be too expensive. Instead, the Vision Zero policy assumed that the only ethically responsible way to deal with traffic safety is to say that no lives should be lost due to road traffic. The policy has been criticised from different perspectives, and its proponents have been accused of being naive, overly ambitious and even unethical. Transport economist Elvik, for example, argues that the aim of eliminating road traffic deaths would be so expensive that other risks would have to be ignored, making Vision Zero counterproductive, given that the more general goal of Vision Zero is to minimise mortality (Elvik, 1999). This would mean, perhaps, that it is both ineffective and unfair.

The Vision Zero policy also introduced a new notion of responsibility that has been expressed as follows:

> The designers of the road transport system are invariably ultimately responsible for the design, management, and use of the system and thus, they are responsible for the level of safety of the system as a whole. The roads users are obliged to abide by the rules that the system designers establish for the use of the road transport system. If the road users fail to abide by the rules – for example due to lack of knowledge, acceptance or ability – or if personal injuries occur, the

> system designers must take additional measures to prevent people from dying or being seriously injured.
>
> (SOU, 2000)

System designers are defined as

> those public and private organizations that are responsible for the design and maintenance of different parts of the road transport system such as roads, vehicles and transportation services as well as those responsible for different support systems for safe road traffic such as rules and regulations, education, surveillance, rescue work, care and rehabilitation.[2]
>
> (Government bill, 1996–1997, p. 17; Nihlén Fahlquist, 2006)

This new approach to responsibility diverges substantially from the traditional view that is also legally codified in many countries, including Sweden, according to which individuals are responsible for traffic accidents. The traditional view does not include shared or collective responsibility, and it does not explicitly ascribe this kind of responsibility to the system designers. Vision Zero has been accused of ignoring or removing individual responsibility, which is allegedly dangerous and wrong. In the libertarian paradigm, Vision Zero represents a wrongful vision of a risk-free society. One might wonder why a risk-free society would be a bad thing. It would be wrong, according to libertarians, not because it would be devoid of risks, but because it would require major government involvement in individuals' lives and infringements on their rights. A risk-free society is one in which the government takes on all responsibility and, in the process, radically infringes on individual freedom (Ekelund, 1999). It would not be a fair society. According to this line of reasoning, people should be allowed to choose their own goals and decide whether the risks associated with these goals are worth taking (Ekelund, 1999). In short, the individual should be allowed to do whatever she decides to do as long as she is not infringing on the rights of other people. This is the so-called harm principle, famously defended by philosopher J.S. Mill (Mill, 1865). We could, for example, imagine that it should be up to the individual to decide whether to use a bicycle helmet when riding a bike in the desolate woods, because an accident there would only affect that individual. However, road traffic is an area where all individuals more or less pose threats to other individuals. It is hard to imagine traffic-related risks that only concern one individual. Even in the case of not using a bicycle helmet, the costs for the government or insurance company to pay for the injuries could be seen as risks, or at least a cost, to others.

Nevertheless, if we focus on the notion of responsibility for road traffic and the idea that system designers are ultimately responsible, we will see that the libertarian criticism is based on a somewhat simplistic view of responsibility in which individual responsibility is *necessarily* reduced if the system designers are ascribed responsibility for road accidents. As should be clear by now, responsibility assignment can be based on more than one value and can have different aims. Consequently, several agents may be ascribed responsibility for the same event or problem and emphasising the responsibility of one agent does not necessarily reduce that of the others.

The distribution of responsibility introduced by Vision Zero asks the questions described above in relation to accidents:

1 Who caused it? (Backward-looking causation.)
2 Who is to blame for it? (Backward-looking blame.)
3 Who should influence/improve things so that it does not happen again? (Forward-looking causation.)

The traditional view of responsibility and traffic safety, focusing as it does on backward-looking causal responsibility and blame, is based on a deontological perspective, emphasising the moral justifiability and fairness of a certain distribution of responsibility. The idea that road users are responsible for accidents has been implicitly or explicitly endorsed, as they were the ones considered causally responsible and sometimes blameworthy. Implicitly, individuals were also seen as being responsible in a forward-looking sense because the solution to the problem of traffic fatalities was assumed to depend largely on the extent to which road users improved their attitudes and behaviour. According to critics, this is the most effective and, more importantly, the only morally acceptable or fair way to ascribe responsibility for traffic safety. Ekelund argues that it is more effective than Vision Zero because too much emphasis on the responsibility of system designers may lead to reckless behaviour by road users (Ekelund, 1999). Furthermore, it is the only fair way of distributing responsibility because individuals should be free to choose their own goals in life and decide which risks are worth taking; in other words, it is their right. Finally, unless individuals are ultimately responsible for risks associated with road traffic, individuals who do take personal responsibility would have to pay for those who do not, and this would be unfair (Ekelund, 1999). Therefore, Vision Zero is allegedly morally unjustified, both from a fairness perspective and from an efficacy perspective.

The new notion introduced by Vision Zero illuminates the value of efficacy, i.e. that a responsibility distribution should contribute to eliminating the risk of fatal road traffic accidents. It was acknowledged in the design of the new policy that both individuals and system designers are causally responsible for accidents. Whereas individuals may be said to trigger accidents through their behaviour, system designers are held causally responsible in terms of underlying systemic conditions.

Whereas individual road users have implicitly been seen as responsible, in a forward-looking manner, in terms of their behaviour, Vision Zero introduces the addition of forward-looking responsibility for the system designers. Although individuals are still considered responsible in a forward-looking sense – for instance, they are expected to "behave responsibly" and follow the rules – the forward-looking responsibility ascribed to the system designers by Vision Zero is substantial. The system designers are given the task of improving the traffic conditions, environment, behaviour and so forth, with the aim of reducing the number of fatalities in the future. It has been argued that forward-looking, or preventive, responsibility has to do with influence and that it is a function of power, impact and knowledge. The more strategic one's position is in these respects, the higher one's responsibility should be (Lenk and Maring, 1991). The Swedish National Road Administration (NRA), which was responsible for traffic safety when Vision Zero was adopted, for example, was expected to have the knowledge, influence and power to improve the roads. The NRA is therefore seen as responsible in a forward-looking sense without necessarily having caused an accident or being to blame for it. The policy-makers designing Vision Zero evidently assumed that the government agency has the power and knowledge necessary to take action to improve the traffic environment and to reduce the number of fatal accidents. If they fail to do what it takes to achieve the desired results, they may be blamed.

The assignment of forward-looking responsibility is not necessarily linked to backward-looking causation or blame. Consider an accident in which a car driver hits a pedestrian crossing a village road at night, resulting in the death of the latter. Although the driver may be considered causally responsible for the death of the pedestrian, the local municipality may be best suited to prevent similar accidents in the future by improving street lighting. As described by former Swedish Director of Road Safety Claes Tingvall, who was also one of the initiators of Vision Zero, the difference between the old and the new approach is that we used to say that 90 per cent of all road accidents are caused by the individual, whereas we now say that 90 per cent of the injuries can be prevented by the system designers (Lind and Schmidt, 2000, p. 22). The implication of this change of focus is that, instead of merely and routinely attributing responsibility to the agent who triggered the unwanted event, we should also consider assigning responsibility to the agent who is able to change the underlying conditions. Thus, what is normally considered to be of secondary importance from a causal point of view becomes a substantial factor in the process of ascribing responsibility.

It might be considered radical to claim that the system designers are ultimately responsible for traffic safety even in cases where individual road users ignore traffic rules and make blatant mistakes. However, it makes sense if this new distribution of responsibility is analysed in terms of the distinction between backward-looking and forward-looking responsibility outlined above. Backward-looking responsibility is still attributed to individual road users. The policy also states that drivers should "take responsibility", i.e. assume forward-looking responsibility to abide by the rules and behave prudently. Furthermore, regardless of the new emphasis on the complexity of causation and the pragmatic question of how the problem of traffic fatalities should be solved, the social practice of blaming is not likely to disappear with the new perspective. There are a lot of cultural, social and political norms and conventions regarding road transportation that will continue to influence the way individuals judge each other's traffic behaviour. Nevertheless, forward-looking responsibility is also explicitly ascribed to the system designers.

This notion is based on the insight that, in order to substantially reduce the number of road traffic fatalities, we have to realise that road users will make mistakes and break the rules. In order to make progress, systems must be designed so that the fatal consequences of such mistakes and the likelihood that certain mistakes and transgressions occur are reduced through regulations and design of the road transport system. Thus, the "ultimate responsibility" is forward-looking responsibility at work. If we want to achieve a safer road traffic environment, we will have to look further than the individual road users who are involved in any given accident. It does not make sense to discontinue the analysis once we know who triggered the accident. Therefore, a forward-looking perspective on responsibility is essential. The way in which we ascribe and distribute responsibility for social problems could be seen as one step in the process towards solving them.

It could be argued that the fact that Vision Zero is not legally binding makes it a "toothless tiger". The question is whether it will have any long-lasting effects if it continues to be just an ethical notion rather than a distribution of responsibility with legal consequences attached. It appears that the goal of Vision Zero will be difficult to achieve without the assignment of forward-looking responsibility to the system designers. However, it is one thing for the government to demand that the relevant agencies do their part and a more difficult task to compel car producers to do their share without enforcement or incentives. It seems clear that it is not enough to have the traffic safety agency do their job, just as it was previously insufficient to only blame individuals for deficient safety. Reaching a level of safety in which the risk of being killed or seriously injured is substantially reduced appears to require a tremendous effort from many different actors, and the question is whether this can be achieved without tangible consequences for system designers who do not assume responsibility. On the other hand, legal sanctions are surely not the only way to effectively make the system designers responsible. For instance, in a market economy, the car industry is vulnerable to pressure from consumers, and negative publicity may be seen as a sanction for not taking responsibility. As we will see, technological design and policy-making concerning its use is one way of implementing forward-looking responsibility for system designers.

Alcohol interlocks – forward-looking responsibility in practice[3]

Driving under the influence of alcohol (DUI) is a serious public health problem, a top contributing factor behind the almost 1.3 million annual deaths in traffic worldwide, as well as many of the more severe injuries (WHO 2004, 2009). Traditionally, drunk driving has been treated as a crime; that is, as a legal responsibility. Due to technological developments such as the alcohol interlock, there is now a way to manage these risks through improvements in technological design, policy-making and an emphasis on forward-looking responsibility. The alcohol interlock is a fairly new technology that promises to eliminate the problem of DUI. The technology has been used extensively in the United States and Canada as a requirement for people convicted of repeated drunk-driving offences (Grill and Nihlén Fahlquist, 2012). This device measures the driver's blood alcohol content (BAC) before the car starts, for example through an exhalation sample. The interlock is connected to the car's ignition, and, if the measured concentration is above the maximum limit, the car will not start. With this device installed in all cars, the risks associated with drink driving could be virtually eradicated.

In 2005, the Swedish government decided that all new cars should have mandatory alcohol interlocks installed by 2012. A report on the technical, economical and legal aspects of universal interlocks was presented by a special commission in the summer of 2006 (Svensson Smith *et al.*, 2006). The bill has not been implemented as of this writing. However, alcohol interlocks are becoming more common in coaches and taxis as a result of decisions made by individual companies. Additionally, since 2012, convicted drunk drivers have been able to get their driver's licence back on the condition that they drive a car that has an interlock installed (www.transportstyrelsen.se/sv/Vag/Alkolas/Alkolas-efter-rattfylleri/). Outside of Sweden, alcohol interlocks are used in many European countries, for example, as a voluntary measure by transport companies, as part of rehabilitation programmes, or in school buses (www.etsc.eu/documents/Drink_Driving_Monitor_July_2011.pdf).

It is important to realise that responsibility for a given event or problem must not be seen as a zero-sum game. For example, making the public health agency responsible for childhood obesity does not necessarily remove responsibility from parents. In certain cases, however, shared responsibility could mean less responsibility for each party. To evaluate the claim that interlocks erode individual responsibility, therefore, we need to thoroughly analyse the case at hand.

The typical response to an accident is to investigate who among those involved caused it and is potentially blameworthy. Interestingly, the narrow focus on individual responsibility can be contrasted with the current dominant perspective in "human factor" research, which is primarily investigated with respect to the aviation industry. The same focus on individual responsibility used to be prevalent in aviation; that is, individual pilots were typically blamed for accidents. Today, research focuses on the context in which pilots make decisions and perform actions (Dekker, 2002). This is appropriate in complex systems like aviation and road traffic where accidents tend to have complex and multiple causes.

When system designers step in to take responsibility for the context in which decisions are made, they may be filling an empty space rather than usurping individual responsibility. As we have seen, the responsibility ascribed to system designers is of the forward-looking kind, aimed at preventing future accidents rather than distributing blame for past accidents. Forward-looking responsibility does include an element of potential blame for future accidents, however, if efforts at prevention turn out to be insufficient. That said, it is essentially a responsibility to get certain things done, rather than to take blame. As we have seen, this should be distinguished from backward-looking responsibility. Importantly, the two kinds of responsibility can co-exist without the one diminishing the other. Indeed, the same kind of responsibility can be ascribed to more than one agent without necessarily diminishing the "total sum" of responsibility. Under the Swedish paradigm, drivers are still expected to do their part in preventing accidents by driving responsibly and following traffic rules. This is also a forward-looking responsibility since failure to drive safely can incur blame even when no accident occurs. A public policy focus in the area of DUI regarding the assignment of backward-looking responsibility to individual road users could, for instance, emphasise incarceration. The main question from that perspective is who is to blame for any given accident. If, on the other hand, focus is on forward-looking responsibility, alcohol interlocks could be a natural way of managing the problem of drunk drivers risking their own lives and the lives of others. This approach involves both system designers taking responsibility by putting an effective systemic solution in place and individual drivers taking responsibility by adjusting to that system.

One question that may be asked is whether technological devices taking over tasks previously performed by people will decrease people's "sense of responsibility"; that is, the moral emotions that make them feel responsible for avoiding harm to others. Are there reasons to believe that ascribing forward-looking responsibility for accident prevention to system designers will in fact make drivers *feel* less responsible for their driving and so less cautious? Technical systems that are very sophisticated and where almost all safety hazards are handled by automatic systems can erode the operator's feeling of responsibility. This has been observed in airplanes, where familiarity with safety devices has led to inattention and complacency on the part of pilots (Perrow, 1999). However, these effects result from safety devices that take over a certain task from the pilot or driver and that work continuously during the whole journey, such as a collision avoidance system. The interlock, on the other hand, merely establishes whether the driver is sober before she can start the engine. This test has no direct effect on the driving experience. It does not at all guarantee that the driver is a good one or that the safety of the driver and of other road users is automatically protected. There are many other safety features and conveniences in cars that do make drivers more passive, such as automatic transmission, cruise control and automatic breaking systems. The interlock, on the other hand, only prevents people above a certain degree of intoxication from driving and is itself passive during the journey. Could it be that, despite these considerations, people will come to think of the interlock as a general test for being fit to drive, such that they will discount the risks of driving tired, stressed or under the influence of other drugs than alcohol? This may of course be possible, since all sorts of misconceptions can spread, but there seems to be no direct reason to expect such a development. It is explicit and obvious that the alcohol interlock measures BAC and nothing else. Could people come to think that activities that are not protected by interlocks are safe to perform after drinking? Again, this seems far-fetched. It is obvious that many activities are risky to perform after drinking and it is common knowledge that drink driving is a serious problem. Attending to this problem should not induce people to forget their everyday experience and knowledge of the impairment that comes with intoxication. In sum, the case for claiming that interlocks erode individual responsibility seems very weak.

The problem of drink driving, and of impaired driving more generally, is a problem where many individuals fail to be responsible enough, with grave aggregate consequences, but where punishment of these individuals is very costly and possibly morally unjustified. The best way to solve such a problem is to change the background circumstances. Directly influencing social norms and increasing the (subjective) probability of detection are two ways to combat the problem, but they are insufficient. Drivers will continue to make mistakes and break the rules. Profound change will only come about through conscious design of the system within which individual decisions and mistakes are made. Today, the technological capabilities of cars provide drivers with opportunities that are illegal and dangerous, such as driving very fast and driving after drinking. The danger is not only to the driver, but to other road users as well. While driving after drinking is not to be dismissed as totally lacking utility, the right to drive after drinking is arguably rather trivial and is far outweighed by other road users' rights to safety. The government should strive to eliminate opportunities that are harmful, dangerous and relatively unimportant from the perspective of individual rights. Eliminating the opportunity to drive after drinking by making interlocks universal, if worth the material costs, seems a perfect example of sound health promotion policy. The alcohol interlock can be seen as a way of implementing the forward-looking responsibility of the system designers, i.e. in this case vehicle designers and policy-makers (Grill and Nihlén Fahlquist, 2012).

Forward-looking responsibility and the distinction between natural and technological risks

The distinction between backward-looking and forward-looking responsibility is highly relevant to technological disasters. Consider the oil spill caused by the Deepwater Horizon explosion in April 2010. Following an explosion in the Gulf of Mexico, there was a major oil spill and 11 people were killed. As is common following disasters like this one, an investigation was initiated, and people wanted to know whether managers and employees at Shell had done something wrong and taken unacceptable risks. The question arose as to whose responsibility it was that people died and the environment was damaged.

Whereas accidents and disasters described as "natural" do not spur the same kind of questions, the occurrence of risk events caused by human beings and technology, like the Deepwater Horizon disaster, usually raise questions about responsibility. One reason could be that human beings enjoy a certain amount of control over the environment (Van de Poel and Nihlén Fahlquist, 2012). Studies in risk perception indicate that people tend to care more about unnatural risks than they do about natural risks. This has been called the "naturalistic fallacy", i.e. our tendency to think that if a risk is natural, we ought to accept it, but if it is not natural it is not acceptable (Hansson, 2004). Even in cases of natural risks – risks with primarily natural rather than human causes – questions about responsibility often appear appropriate. It is very difficult to demarcate the "natural" from the "unnatural", the "artificial", or the "human-made". Although it may be far-fetched to hold someone responsible for, for example, an earthquake killing thousands of people, it may be reasonable to hold certain people responsible for the lack of a proper warning system for earthquakes. Furthermore, if there was a nuclear facility placed in an area where there is a substantial risk that earthquakes may occur, it may additionally be reasonable to ascribe responsibility to someone for that decision (Van de Poel and Nihlén Fahlquist, 2012). As these examples illustrate, natural and man-made risks are not all that easy to distinguish. Nevertheless, we often refer to risks as either man-made or natural.

The concept of naturalness is relevant in different ways in the context of risk. As we saw above, "natural" risks are sometimes seen as unavoidable and something for which no one is responsible. In other contexts where risks are discussed, "natural" is seen as a beneficial advantageous adjective. "Natural" is what philosophers call a "thick concept", combining non-evaluative and evaluative features. Similarly, vaccine opponents sometimes refer to the naturalness of not vaccinating and instead letting children be exposed to bacteria and viruses, thereby presumably developing a better immune system. These lines of argument can, have, and should be questioned. The point here is merely to illustrate how the concept of naturalness is value-laden and frequent in societal debate, its connotations being both positive and negative depending on the context.

The difficulties of distinguishing what is natural has made some theorists question the very distinction between natural and artificial. For example, Hansson argues that the distinction between what is natural and unnatural more or less dissolves when logically scrutinised (Hansson, 2004). Many things that we take for granted as good, such as glasses, are not "natural" but rather so common that we think of them as more or less natural. In addition, other things that are natural are highly toxic and dangerous and hence we should avoid them. However, the distinction between "natural" and "artificial" is central to environmental philosophy. To protect the environment has often been taken to mean protecting the nature that is unaffected by human activities. In 1989, McKibben famously argued that there is no longer any nature that is unaffected by human beings (McKibben, 1989). Thus, there is only "artificial" environment. That said, as Vogel argues, the end of nature does not necessarily entail the end of environmental concerns. Even though it is difficult to demarcate what is natural and what is artificial, we should protect the *environment*, whether natural or artificial, against harm (Vogel, 2016).

Even though it is difficult, or impossible, to say exactly what is natural and what is not natural in relation to some risks, it is reasonable to require that if we, as human beings, bring new risks into the world, we should be accountable if or when these risks materialise. After all, it is possible to distinguish between the earthquake on the one hand and the decision to place a nuclear plant in the area on the other. Consequently, if a company introduces new risks when introducing a new technology, it should arguably be held accountable; that is, it should be ascribed backward-looking responsibility if these risks materialise. In cases where the causes are not so clearly human activities, it is less clear if anyone, and, if so, whom, should be held accountable.

Consider these two scenarios:

1 Five people die as a consequence of a hurricane.
2 Five people die because an oil rig explodes.

It is one thing to argue that we all, as a society, have a responsibility to make sure that the effects of climate change on people's lives and the environment are kept to a minimum. However, is it not more condemnable, and has not a more serious wrong been done in the second case? Arguably, bringing more risks into an already risky world ought to require more responsibility-taking. Even though it is important to be able to blame those who introduce risks in an ethically illegitimate way, it is even more important to make sure that such introduction of risk is prevented or minimised. Arguably, what is needed today is the cultivation of a sense of forward-looking responsibility within groups of professionals whose activities introduce new risks into society. This notion is based on virtue ethics, which is discussed in Part II.

One might ask whether this notion would lead us to adopt the precautionary principle, i.e. to always opt for the least risky option. However, as has been argued, to do nothing is not an option without consequences either (e.g. Sandin, 2004). Given the current level of innovation, society faces a dilemma: society needs progress, and yet progress inevitably entails risks to human health and environment. Furthermore, there is significant uncertainty. Therefore, introducing new technologies in society, in a sense, inevitably entails a social experiment (Van de Poel, 2016). The challenge is to make sure, as far as possible, that technological development and the social experiments it involves are achieved as responsibly as possible. Therefore, individuals taking part in activities through which risks are created ought to take forward-looking responsibility.

Conclusion

Against the background of our analysis thus far, it is clear that responsibility assignment and distribution can be evaluated in terms of fairness and efficacy. In addition, they can be of two primary kinds: backward-looking or forward-looking. That gives us a general matrix for analysis and evaluation of responsibility assignment and distribution. Returning to the question raised in the previous chapter concerning cancer prevention and responsibility, we could now make the question more complex:

1 Is it fair to hold individuals responsible for having developed cancer? (**Backward-looking and fairness perspective.**)
2 Is it fair to ascribe forward-looking responsibility to individuals, i.e. to communicate the need for them to take measures to prevent cancer from developing? (**Forward-looking and fairness perspective.**)
3 Is it effective to hold individuals responsible for having developed cancer? (**Backward-looking and efficacy perspective.**)
4 Is it effective to ascribe forward-looking responsibility; that is, to take measures to prevent cancer from developing? (**Forward-looking and efficacy perspective.**)

When analysing and discussing a particular distribution of responsibility, such as the one introduced by Vision Zero, we could use the following simple matrix as a starting point.

Table 2.1 Responsibility: core values and kind

Responsibility – kind *Responsibility – value*	*Backward-looking responsibility*	*Forward-looking responsibility*
Fairness	1	2
Efficacy	3	4

Notes

1 This section is largely based on the article Nihlén Fahlquist, J. 2006. Responsibility ascriptions and Vision Zero. *Accident Analysis and Prevention* 38: 1113–1118.
2 All translations are the author's unless otherwise specified.
3 This section is based on the author's contribution to Grill and Nihlén Fahlquist 2012. The author has asked the permission of the co-author to use this part of the published article.

References

Bovens, M. 1998. *The Quest for Responsibility. Accountability and Citizenship in Complex Organisations*. Cambridge: Cambridge University Press.

Dekker, S. 2002. *The Field Guide to Human Error Investigations*. Aldershot: Ashgate.

Downie, R.S. 1971. *Roles and Values: An Introduction to Social Ethics*. London: Methuen.

Dworkin, G. 1981. Voluntary health risks and public policy: taking risks, assessing responsibility. *Hastings Center Report* 81(11): 26–31.

Ekelund, M. 1999. *Varning – Livet kan leda till döden! En kritik av nollvisioner*. Stockholm: Timbro.

Elvik, R. 1999. Can injury prevention efforts go too far? Reflections on some possible implications of Vision Zero for road accident fatalities. *Accident Analysis and Prevention* 31(3): 265–286.

Feinberg, J. 1970. *Doing and Deserving: Essays in the Theory of Responsibility*. Princeton, NJ: Princeton University Press.

Gallie, W.B. 1955. Essentially contested concepts. *Proceedings of the Aristotelian Society* 56: 167–198.

Garvey, J. 2008. *The Ethics of Climate Change. Right and Wrong in a Warming World*. London: Continuum.

Goodin, R.E. 1986. Responsibilities. *Philosophical Quarterly* 36: 50–56.

Grill, K. and Nihlén Fahlquist, J. 2012. Responsibility, paternalism and alcohol interlocks. *Public Health Ethics* 5(2): 116–127.

Hansson, S.O. 2004. Fallacies of risk. *Journal of Risk Research* 7(3): 353–360.

Husak, D. 2004. Vehicles and crashes: why is this moral issue overlooked? *Social Theory and Practice* 30(3): 351–370.

Johansson, L. 2010. The functional morality of robots. *International Journal of Technoethics* 1(4): 65–73.

Lenk, H. and Maring, M. 1991. A pie-model of moral responsibility? Remarks concerning ethical dilutionism. In: G. Schurz and G.J.W. Dorn (eds), *Advances in Scientific Philosophy: Essays in Honour of Paul Weingartner on the Occasion of the 60th Anniversary of His Birthday* (pp. 483–494). Amsterdam: Rodopi.

Lind, G. and Schmidt, K. 2000. Leder nollvisionen till det trafiksäkra samhället? En kritisk analys av effektiviteten i trafisäkehetsarbetet. *KFB* Nr 1.

Lindgren, A. 1978. *Känner du Pippi Långstrump?: bilderbok*. 7. uppl., 3. tr. Stockholm: Rabén & Sjögren.

McKibben, B. 1989. *The End of Nature*. New York: Anchor.

Mill, J.S. 1865. *On Liberty*. London: Longman.

Mullin, A. 2007. Children, autonomy, and care. *Journal of Social Philosophy* 38: 536–553.

Nihlén Fahlquist, J. 2006. Responsibility ascriptions and Vision Zero. *Accident Analysis and Prevention* 38: 1113–1118.

Nihlén Fahlquist, J. 2009. Moral responsibility for environmental problems – Individual or institutional? *Journal of Agricultural and Environmental Ethics* 22(2): 109–124.

Nihlén Fahlquist, J. 2017a. Responsibility analysis. In S.O. Hansson (ed.), *The Ethics of Technology: Methods and Approaches*. London: Rowman & Littlefield.

Nihlén Fahlquist, J. 2017b. The gap of institutional responsibility for environmental problems. In T. Scavenius and S. Rayner (eds), *Institutional Capacity for Climate Change Response: A new Approach to Climate Politics*. Abingdon: Routledge, 2018.

Peden, M., Scurfield, R., Sleet, D., Mohan, D., Hyder, A.A., Jarawan, E. and Mathers, C. (eds). 2004. World report on road traffic injury prevention. Geneva: World Health Organization.

Perrow, C. 1999. *Normal Accidents*. Princeton, NJ: Princeton University Press.

Roeser, S. 2006. The role of emotions in judging the moral acceptability of risk. *Safety Science* 44: 689–700.

Roeser, S. 2010. *Moral Emotions and Intuitions*. Basingstoke: Palgrave Macmillan.

Richardson, H.S. 1999. Institutionally divided moral responsibility. In: E.F. Paul, F.D. Miller and J. Paul (eds), *Responsibility* (pp. 218–249). Cambridge: Cambridge University Press.

Sandin, P. 2004. *Better Safe than Sorry: Applying Philosophical Methods to the Debate on Risk and the Precautionary Principle. Diss.* (sammanfattning) Stockholm: Tekn. Högsk.

Slovic, P. 2000. *The Perception of Risk*. London: Earthscan.

Smiley, M. 1992. *Moral Responsibility and the Boundaries of Community: Power and Accountability from a Pragmatic Point of View*. Chicago, IL: University of Chicago Press.

SOU. 2000. Ett gemensamt ansvar för trafiksäkerheten. Betänkande av Trafikansvarsutredningen (report by governmental committee) Fritzes offentliga publikationer, Stockholm, p. 43 (in Swedish).

Svensson Smith, K., Nilsson, M. and Schönning, O. 2006. *Öppna möjligheter med alkolås – Slutbetänkande av Alkolåsutredningen* (Open possibilities with alcohol interlocks – Final report of the Commission on Alcohol Interlocks). Stockholm: SOU 2006: 72.

Van de Poel, I. 2016. An ethical framework for evaluating experimental technology science and engineering. *Ethics* 22(3): 667–686.

Van de Poel, I. and Nihlén Fahlquist J. 2012. Risk and responsibility. In S. Roeser, R. Hillerbrand, M. Peterson and P. Sandin (eds), *Handbook of Risk Theory*. Dordrecht: Springer.

Van de Poel, I., Nihlén Fahlquist, J., de Lima, T., Doorn, N., Royakkers, L. and Zwart, S. 2012. The problem of many hands: climate change as an example. *Science and Engineering Ethics*, 18: 49–67.

Vogel, S. 2016. *Thinking Like a Mall: Environmental Philosophy After the End of Nature*. Cambridge, MA: MIT Press.

WHO. 2004. World report on road traffic injury prevention. Geneva: World Health Organization.

WHO. 2009. *Global Status Report on Road Safety: Time for Action*. Geneva. World Health Organization.

WHO. 2017. Number of traffic road deaths. Geneva: World Health Organization. www.who.int/gho/road_safety/mortality/number_text/en/ (accessed 7 December 2017).

WHO. 2018. Global Health Observatory (GBH) Data. Geneva: World Health Organization www.who.int/gho/road_safety/mortality/number_text/en/ (accessed 20 Febuary 2018).

Young, I.M. 2006. Responsibility and global justice: a social connection model. *Social Philosophy and Policy* 23(1): 102–130.

3 Individual and collective responsibility

Introduction

As we have seen, responsibility can be evaluated in terms of the values of fairness and efficacy, and it can be backward-looking or forward-looking. We will now turn our attention to a third distinction, which is useful when analysing responsibility for problems related to risks. Most of the topics discussed in society, which focus on responsibility, involve a tension between individual actions and societal activities. Consequently, they can be framed in terms of individual or collective responsibility, or a combination of these. Discussions on these topics are usually politically and ideologically heated. What is actually being discussed could perhaps become clearer if the distinctions discussed in the previous chapters were used. If so, it would become explicit that what we are really discussing is whether it is *fair* to ascribe responsibility for, e.g. health, to the individual or to the government, and what would be the most *effective* way to distribute responsibility for health. In addition, individual and collective responsibility for social issues such as health or climate change could either be discussed in terms of backward-looking or forward-looking responsibility.

The notion of collective responsibility

In modern society it is common to talk about collective entities, e.g. governments, nations and companies, as being responsible. As we saw in the case of Swedish traffic safety policy, Vision Zero attributes moral responsibility to both public and private system designers. In an everyday context, it may appear simple enough, but philosophically, collective moral responsibility is a contested notion.

Consistent with the notion of individualism that is so well-entrenched in liberal Western societies, many philosophers traditionally regard the notion of collective responsibility with suspicion. This is hardly surprising given what was done in the rhetorical context of collective group identity during the twentieth century.

After World War II, many people blamed the German people for the atrocities committed during the Holocaust. In 1948, this notion was philosophically challenged. In a famous article, Lewis argues that only primitive peoples treat the collective, such as the family, nation, or tribe, as the main unit. He argues that ideas of collective responsibility are "simply the obverse of the tendency to set some abstract good of the community above the wellbeing of its individual members" and that this notion is closely associated with oppressive and totalitarian ideas and practices (Lewis, 1948). According to Lewis, only individuals can have values and be morally responsible. The only legitimate basis for moral blame is what individuals intentionally bring about, and the alternative view would entail a substantial risk of falling back into a barbarous, tribal way of thinking. No one, Lewis argues, should be held responsible for the conduct of another (Lewis, 1948). Using the distinctions we have discussed in previous chapters, it is noteworthy that Lewis discusses moral responsibility as blameworthiness, i.e. responsibility in the backward-looking sense. He states that, although no one can be said to be responsible in the sense of being answerable for the conduct of others, we all have duties towards our fellow human beings. This view could be interpreted as denouncing collective backward-looking responsibility while allowing for collective forward-looking responsibility.

Whereas Lewis did not consider it fair, or morally legitimate, to ascribe moral responsibility to collective entities, others take a more pragmatic outlook. Cooper argues that we *do* in fact ascribe responsibility to collectives, e.g. nations. This fact forces us to choose whether to (1) claim that this is a misuse of the concept of responsibility, (2) state that ascriptions of collective responsibility are reducible to individual responsibility or (3) argue that collective responsibility is not reducible to individual responsibility, but that it is different than and additional to individual responsibility (Cooper, 1968).

If we agree with the notion that there is collective responsibility, then we must ask under what circumstances a collective entity can be considered to be the kind of entity that could reasonably be held responsible. French, who defends the idea of collective responsibility, distinguishes between *aggregate collectivity*, which is the concept for a simple collection of people, and *conglomerate collectivity*, which is an "organization of individuals such that its identity is not exhausted by the conjunction of the identities of the persons in the organization" (French, 1987, p. 13).

If a so-called aggregate collectivity is held responsible, this responsibility ascription is reducible to the constituent individuals, he argues. However, when responsibility is ascribed to a conglomerate collectivity, it is not. This means that when an entity is a conglomerate collectivity, it is to be treated as a moral agent, in which case the whole is considered to be greater than the sum of its parts (French, 1987). This account of collective moral responsibility has been criticised because of its allegedly absurd implications. Critics have argued that treating corporations as moral subjects entails that they also receive the value and rights of individual moral persons, which would be morally wrong (Pfeiffer, 1988).

As has been noted by some theorists, there is a discrepancy in Western societies between the ideal of autonomous individuals on the one hand and the prevalence of large bureaucratic organisations on the other (Bovens, 1998). These "complex organisations" are large-scale and have a bureaucratic structure and formal status, i.e. they can sign agreements and contracts. In the twentieth century, these entities grew in number and their social importance increased. Thus, they have become "corporate actors" (ibid.). While they have been granted legal status that is almost equal to that of natural persons, they are obviously more powerful (ibid.). In fact, a remarkably large proportion of social activities take place in the context of private as well as public hierarchical organisations. Technological development, with beneficial as well as negative effects, also occurs in this context. In a contemporary risk society, corporate actors create previously unimaginable opportunities for people and for society, but they also introduce substantial risks to health and environment. Thus, collective agency has de facto increased, but the notion that individuals are the only ones who can be autonomous and responsible is still strong in theory.

Given the substantially increased social role of corporate actors, it seems extremely important to move beyond the traditional debate on moral responsibility as merely concerning the relationship between the individual agent and her actions and omissions. It is highly problematic to claim that only individuals can be morally responsible while complex organisations have the power to completely change the lives of human beings to the extent that we are in fact dependent on them (Bovens, 1998). As we discussed in the previous chapter, capacity and power arguably entail responsibility. In feudal societies, relationships between individuals were clearly unequal. What is new in the asymmetrical social structure of today is that the unequal relationship is not between individuals, but between two "fundamentally different types of actors: people and complex organisations" (Bovens, 1998, p. 19).

The activities of complex organisations can and do introduce risks into society. The risks and uncertainties of today are different both quantitatively and qualitatively than the risks of previous times. In fact, risks associated with the activities of corporate actors can affect large groups of individuals to the extent that society is disrupted, or its structure fundamentally altered (Bovens, 1998).

Thus, considering the new role of complex organisations as well as the power of technological innovation and development and the capacity of collective agents, it would be highly problematic to exclude the notion of collective responsibility from an analysis of responsibility and risk. Instead of questions like "What is the nature of the relationship between the agent and her actions?" and "When is an action intended?" we should be asking

"How can we arrive at a reasonable distribution of responsibility so that complex organisations, public as well as private, act to achieve optimum use of human resources and to avoid exposing human beings and society to unreasonable risks?" Private companies, governments and other organisations may not act or intend in the same way as individuals, they may not *feel* guilty or fear sanctions like humans. However, the power, risk-inducing potential and capacity of modern collective agencies necessitates a notion of collective moral responsibility.

One way forward is to hold them accountable through laws and law-enforcing mechanisms, through liability and backward-looking responsibility. This is, to some extent, being done. Another way would be to foster an environment where responsibility as virtue is promoted, i.e. for example a sense of ownership and not merely leaving things to others, actively taking action and caring about other people.

We will have a closer look at the concept of responsibility as virtue in relation to collective responsibility in Part II, especially in the chapter analysing technological risks. In the remainder of this chapter, we will explore a context in which the tension between individual and collective responsibility is highly salient, i.e. environmental problems.

Environmental responsibility – individual or collective?[1]

One issue that has become a major topic for societal debate in the last decade or two is environmental responsibility. Especially in the years preceding the financial crisis of 2008, the question of individual responsibility for climate change was the subject of much debate. This discussion will illustrate the tension between the notion of individual responsibility and the notion of collective responsibility that is inherent in contemporary debates concerning risk management. The same tension exists in the debate about personal versus collective responsibility for health.

The environmental debate during the early twenty-first century focused on the following question: To what extent are environmental problems the responsibility of individuals, as consumers and citizens? With the knowledge we have today about the causes of environmental problems and the fact that citizens in many industrialised societies are well-informed about their own role in contributing to the problems, individuals appear to have some responsibility. People choose to behave in ways that contribute to the problems or to their solutions. Hence, as individuals we are taken to be morally responsible for environmental problems. It has been pointed out that the notion of the consumer as an active and responsible agent is embraced by strong actors within governments, the corporate sector and NGOs. Furthermore, it is believed that we can make a difference when making choices about what to buy, and the individual consumer is often seen not only as responsible for herself, but directly responsible for the world. As Micheletti (2003) argues, "everyday choices and acts by individuals play an important role for the future of political, social, and economic life. In short, every person is part of global responsibility-taking". However, Jacobsen and Dulsrud (2007), criticise this idea of a "generic active consumer" and claim that this is far from a universal entity. They argue that consumers appear to be strongly influenced by cultural, social and institutional settings and thus have very different attitudes and beliefs depending on where they live. This means that the voluntariness with which individuals, as consumers and citizens, act is not as substantial as is sometimes assumed. As we saw in the previous chapter, voluntariness, i.e. freedom and knowledge in the Aristotelian framework, is a condition for responsibility as blameworthiness.

These insights should be used in our discussion of individual moral responsibility for environmental problems like climate change. As we have seen, a common assumption is that we are responsible to the extent we causally contributed to a problem. This notion has previously led to the view that because my actions are so marginal, I am not at all responsible for the overall environmental problem, e.g. air pollution, the problem of waste, or climate change. This, I believe, is not correct. However, replacing the idea that individuals are exempt from responsibility for environmental problems with a notion of full individual responsibility is too hasty. I will argue that the extent to which individuals should be held responsible for environmental problems needs to be placed in context. More specifically, I will argue that an individual should not necessarily be held responsible in the backward-looking sense. This means that she should not be blamed for taking less environmentally friendly actions instead of more environmentally friendly actions, if she lacked reasonable alternatives or if the cost of choosing the right action was too high. The individual's choice to act in a certain way should be considered in her socio-economic, political and cultural context. As we have seen, the idea that agents should be excused in circumstances when their actions were not fully voluntary has been asserted by philosophers since Aristotle. However, this reasonable idea appears not to be taken fully into account in the current public debate on individual responsibility for environmental problems. Although it is promising that we are now starting to publicly debate environmental problems to a previously unseen extent and to acknowledge the role of human beings, we now seem to run the risk of putting too much focus on the behaviour of individuals at the expense of illuminating the vital role of collective agents.

Although sometimes excused from backward-looking responsibility, I will argue that some individuals should be ascribed forward-looking responsibility. Individuals who have reasonable alternatives, capacity and resources to do something about the environment should be seen as responsible. Collective agents have the power and resources to affect the number of individuals who possess such capacity and resources. Because of this, a great share of responsibility should be assigned to governments and corporations. As we will see, such collective actors are responsible (in a forward-looking sense) because (a) they have power and resources to do more to solve environmental problems, and (b) they have the capacity to make it easier and less costly for individuals to act in environmentally friendly ways.

Before moving on, let us make two conceptual clarifications. First, when I discuss individual responsibility in what follows, I primarily refer to individuals as individual consumers and/or citizens. Second, when I discuss institutions or institutional agents, I primarily refer to governments and corporations, but the discussion is also relevant to local communities and NGOs to some extent.

The analysis will begin by looking at some examples of cases where questions concerning individual responsibility for environmental problems can be, and frequently are, asked. Second, the concept of responsibility in the environmental context is discussed. Third, the notion that individuals are blameworthy for performing acts that contribute to environmental problems is criticised. Fourth, the suggestion that individuals are responsible in a forward-looking sense is presented. Fifth, the forward-looking responsibility of institutional agents is discussed.

Examples: the intuitive level

Imagine Agnes, a citizen in a modern, well-ordered society. In Agnes's residential area, which is socio-economically underprivileged, there is no recycling station, and Agnes, who cannot afford to drive a car, would have to travel by public transport to get to the nearest recycling station. Agnes is a single mother, and it is practically very challenging for her to take her waste to the recycling station. However, not recycling her waste increases the risk of not achieving a sustainable society. Each time she, as an individual, refrains from recycling, her contribution to environmental degradation is marginal, almost undetectable, but if she does it on a regular basis – and if other people do it – the consequences are unwanted and more severe. Her options are as follows.

1. Going to the recycling station and putting her recyclable waste in the appropriate containers. In this case, she has three options:
    a. Bringing her three children, paying for bus tickets, and going to the recycling station.
    b. Hiring a babysitter, paying for a bus ticket, and taking the bus to the recycling station.
    c. Borrowing someone's car to go to the recycling station.
2. Throwing the cans in the household trash.

If she lived in a place where the recycling station was merely a short walk or if she had someone who could help her by taking her waste to the recycling station or looking after her children while she went there, that would have made a difference in terms of her options. In this case, as it is described above, to what extent is Agnes responsible for the problem of non-recycled waste? Intuitively, given the difficulties described, it does not appear reasonable to hold Agnes fully responsible for choosing the second option, that is, for throwing the cans in the household trash.

A second example relates to the suggestion that consumers should choose "climate-smart" food over food that has been transported long distances by airplane or truck or which required energy-intense production. Imagine Bernard, a consumer who refrains from buying "climate-smart" food. Bernard is economically under-privileged, and climate-smart vegetables are more expensive than non–climate-smart vegetables. Buying climate-smart vegetables would mean that he could not afford a monthly bus pass, which he needs in order to commute to work every day. When compared to another individual such as Charles, who lives in the same society but who is well-off, there is obviously a crucial difference between Bernard and Charles in terms of the cost of choosing the environmentally friendly action.

A third example is the notion that we should all contribute to the reduction of carbon dioxide emissions by using public transport instead of driving our own cars. The environmentally friendly thing to do is to take public transport instead of driving a car to work. By choosing our own car instead of taking the bus or train, we marginally increase the risks of climate change. However, many societies lack a reliable, safe and adequately extensive public transport system, and this makes a relevant difference in terms of what can justifiably be required. Imagine an individual, Donna, who lives in a society where public transport is well-developed, and another individual, Emma, who does not. The crucial difference between Donna and Emma lies in the options open to them. It is not asking too much to expect Donna to use public transport, thereby contributing to a better environment or at least not polluting it. However, expecting Emma to do the same is asking a lot more, since the alternative is either not there or is there but the cost of choosing it (in terms of money, time and/or convenience) is simply too high. Whereas we still believe she should do what she can to use public transport instead of driving a car, we should not put the same kind of demands on her as we do on Donna. The alternatives presented to Emma may be the following:

1 Driving for one hour.
2 Walking five minutes to the subway station and taking the subway to work, which is a one-hour journey.

Compare this to Emma's options below.

1 Driving for one hour.
2 Walking 1,000 metres to the bus stop and waiting for the bus, which is normally not on time, and then travelling for two hours on the bus.

We have now looked at some cases in which an individual's environmentally damaging act is put in the context of the options she had when acting. In all of the cases, if many people choose the environmentally damaging thing, the risks of climate change and unsustainability will increase. However, one person acting only marginally increases the risks. In the first case the problem lies in the practical challenges or the effort that has to be made in order to do the environmentally friendly thing. In the second example, the obstacle is money. In the third example it is the infrastructure that is different in Donna's and Emma's cases. The infrastructure is arguably the result of political decisions, culture and tradition, as well as economic factors. The examples are common in everyday life, and I think many of us, if confronted with such examples, would agree that people have different sets of alternatives when acting due to their different contexts and that the choice to act in environmentally damaging ways can be more or less voluntary. I argue that this should be taken into account when discussing individual responsibility for environmental problems. I believe many of us share this intuition, and many philosophers have theorised about the importance of alternatives for acting and the importance of voluntariness to moral responsibility. Yet this notion is frequently ignored in the public debate on individual responsibility for the environment, where all individuals appear to be considered as equally appropriate subjects of responsibility, as if either all individuals are responsible for the environment or no individuals are responsible for the environment. Before continuing the discussion, we should make some clarifications regarding the concept and notions of moral responsibility. As we have seen in previous chapters, the concept of moral responsibility is complex and is used differently in different contexts.

The notion of responsibility I question in the following chapter is backward-looking, referring to actions performed by an individual in the past that are believed to have caused negative outcomes and where the agent is considered blameworthy. This idea will be contrasted with a forward-looking notion of responsibility.

So, what is it to be responsible in the backward-looking sense? As we saw in Chapter 1, Strawson argues that moral responsibility consists of those reactive attitudes, for example resentment and gratitude, we hold towards each other as co-members of the moral community (Strawson, 1962). In our present context this could mean, for example, that when someone who is concerned about the environment sees Agnes throwing aluminium cans in the trash bin, she is likely to react with some resentment and to hold Agnes responsible. She might blame her for performing the action that contributes to the problem of waste instead of the action that is more environmentally friendly, i.e. to recycle the cans.

However, as we saw in the introduction to Part I, the notion of reactive attitudes appears not to tell the whole story of what moral responsibility is. As members of the moral community, we also have the ability to step back and analyse situations from a more detached perspective, and most of us are also able to assess and alter our conduct, for example when confronted with new information. When the person blaming Agnes learns more about Agnes's situation, she may analyse the situation differently and change her view about Agnes's responsibility.

It has been argued that Strawson's view makes it difficult to criticise practices of holding agents responsible and that it is, to some extent, static and relativistic. There is a difference between being *held* responsible and *being* responsible that is not captured by the theory of responsibility as reactive attitudes. To be morally responsible is not merely to be a target of reactive attitudes, as Strawson argues. According to Fischer and Ravizza's revised Strawsonian theory, "agents are morally responsible if and only if they are appropriate recipients of reactive attitudes" (Fischer and Ravizza, 1993). Hence, whereas the reactive attitudes towards the actions of co-agents in a moral community is one element of moral responsibility, another is critical reflection about these practices. Using these notions, the idea that I would like to reflect critically upon is the following:

> Individuals, as consumers and/or citizens, are appropriate recipients of reactive attitudes, e.g. blame, when acting in ways that, in the aggregate, contribute to environmental problems.

Individuals and blameworthiness

For most of the twentieth century, individuals were not expected to do anything for the environment; instead these problems, if recognised, were merely discussed on an international level by governments as a matter of legislation and policy. This approach, I think, was too extreme in its refusal to acknowledge individual responsibility. It was too extreme for two reasons. First, if we are to solve environmental problems, it will in the end partly require people to act in ways that promote such solutions. The question is how to get there, not *whether* we should get there. Second, a society in which people care about the environment, perhaps by developing green virtues (see Jamieson, 2007), is likely to be a better society. It appears likely that a society where people care about the environment is better than a society where people do not care about the environment. We will analyse virtues and their relation to responsibility in Part II. In the following section, I will focus on the risk of over-emphasising individual responsibility in the context of environmental problems. I do not disagree with those who claim that it is ultimately a matter of individuals behaving in ways that promote a better environment. I wish to question the leap from that observation to the conclusion that individuals are blameworthy for acts that contribute, in the aggregate, to environmental problems.

In the beginning of the twenty-first century, climate change came to the fore as the major global challenge, and it started to be discussed as a moral issue. Individuals were increasingly seen as the responsible agents and appropriate targets of blame when they refrain from acting in climate-smart ways such as recycling, turning the thermostat down, driving a car instead of using public transport or cycling and so forth. Accordingly, demands that individuals' choices regarding, for example, consumption and transportation should be morally scrutinised were expressed in public debates. There were, for instance, tests on various websites where one could answer a set of questions to see how "climate-smart" one lives, whether one is "environmentally aware" and so forth. Consumers who bought organic food started to be called "ethical consumers". The implication of this new notion of full individual responsibility for the environment appeared to be that if one does not always choose the most environmentally friendly option, one is, to some extent, an "unethical" or at least a "non-ethical" or "irresponsible" consumer, hence an appropriate target for blame.

The primary problem with the question of whether individuals should be held responsible when acting in ways that contribute to environmental problems is that it is often assumed to be a choice between the following statements:

1 Individuals *should be held responsible* when acting in ways that contribute to environmental problems.
2 Individuals *should not be held responsible* when acting in ways that contribute to environmental problems.

The nuances and complexity of the matter are ignored. What is not acknowledged when posing the question as a choice between 1 and 2 is that individuals are different and exist in different socio-economic, political and cultural contexts. Consequently, it is not necessarily the case that *all* individuals are *always* responsible or that individuals are *never* responsible. Individuals are arguably responsible, that is, are appropriate targets of blame, for some of their environmentally unfriendly actions and not for others. The notion that individuals perform such acts with different degrees of voluntariness should inform our ascriptions of responsibility.

As we saw in previous chapters, the notion that the degree of voluntariness with which an agent performed an act affects the blameworthiness of that agent has been present in philosophical discussions since Antiquity. According to Aristotle, agents are only blameworthy for voluntary acts. Recall that voluntary acts are acts not performed under (i) compulsion or (ii) due to ignorance (Aristotle, 2000). Whereas individual responsibility (in the backward-looking sense) was for a long time not discussed in debates about environmental problems, individuals are increasingly expected to behave in certain ways because of changing norms. Individuals in modern, well-ordered societies are now expected to buy more organic food, to use public transport, to ride a bike, to lower their thermostats and so forth. However, if there is too much focus on individual behaviour and individual responsibility, we risk losing sight of social, economic, cultural and political contexts. Jacobsen and Dulsrud rightly criticise this idea of a "generic active consumer" and argue that this individual is far from a universal entity (Jacobsen and Dulsrud, 2007). Instead, consumers appear to be strongly influenced by cultural, social and institutional settings and thus have very different attitudes and beliefs depending on where they live. However, focusing too much on individual responsibility is not problematic just because individual differences are not acknowledged to an adequate extent. It is also problematic in terms of fairness, and because there are often structural and societal features that work as obstacles to most citizens in a society.

What I am arguing here is the following. Individuals are blameworthy for acts that contribute to environmental problems, e.g. refraining from recycling, driving instead of using public transport or a bicycle, when it is reasonable to expect them to choose the environmentally friendly option. This means that when there are reasonable alternatives to act in a more environmentally friendly way or when the cost and effort of performing the environmentally friendly act is reasonable, an individual is potentially blameworthy for not choosing the right option. This is so because when they had that alternative, for example when there is a highly developed and accessible recycling system, public transport system, or bicycle lanes, it is reasonable to demand that they choose the environmentally friendly option. Similarly, if the individual is well-off, it is reasonable to expect her to buy climate-smart food even if it is expensive. If such opportunities to do right, to act in environmentally friendly ways exist to an adequate extent, then there are no excuses and individuals should be expected to do so. However, if an agent is poor and climate-smart food is expensive, then it is not fair to hold her responsible for buying the less expensive and less environmentally friendly alternative.

Reasonable alternatives

What, then, can be said about the reasonableness of alternatives? When does an individual consumer or citizen have reasonable alternatives? As mentioned in Chapter 2, Aristotle argued that an individual should be excused from (backward-looking, in my terms) responsibility (blame) when she did not act voluntarily, i.e. acted under compulsion or due to ignorance.

The lack of reasonable alternatives can be stated in similar terms. Although consumers are not coerced into buying food that was produced using unnecessarily high levels of energy, if organic or climate-smart food is very expensive compared to regular food, the alternative to buy the environmentally less damaging food is not really an option unless the consumer is wealthy. This is an example of how the cost of choosing the "right" option is highly relevant. In addition to cost, the availability of good options is relevant. If 99 per cent of the products in grocery stores are not climate-smart products, this is obviously an obstacle to people who want to reduce their carbon footprint even if they can afford to buy those products. Similarly, unless there are safe and extensive bicycle lanes, the alternative to ride a bike instead of driving a car is not a realistic option for people who drive to work. The latter is an example of how culture, tradition and political decisions affect how people choose to act. They may choose to drive their car because they grew up and live in a society that treats cars and highways as very important parts of life, culture and society. On the other hand, they may resist habits and cultural pressure and choose to ride a bicycle or use public transport instead. However, if the infrastructure makes it difficult or *too* difficult or inconvenient to choose those options, the conscientious citizen may, very reluctantly, continue to drive her car to work. In real life, compulsion appears to be a matter of degree. It can be more or less difficult to choose a certain behaviour, depending on circumstances.

It is obviously difficult to draw a line between what is reasonable and what is not in individual cases concerning what should be expected. On the political and institutional levels, it is clear that some infrastructural, cultural, socio-economic and political features of modern societies are questionable from an environmental perspective, and that these features counteract individually sustainable behaviour. The most obvious problem is likely to be the car dependency of industrialised nations. A second problem may be the extent to which people in Western societies are used to eating meat products. A third problem is the overuse of packaging and plastic bags.

In addition to cost and the availability of good options, information about the environmental footprints we leave is obviously crucial. However, the question of available information is not as straightforward as one might think. In most well-ordered societies the information is available, which means that most people have the opportunity to know how their behaviour affects the environment. However, even if there is a lot of information, it is sometimes too expensive to do the environmentally right thing. Furthermore, in many cases, the information does not reach all groups of people in society.

All of these factors – cost, convenience, availability and access to information – affect the reasonableness of the alternatives individual consumers and citizens have when they "choose" to act in environmentally friendly or environmentally unfriendly ways. Because these are highly prevalent problems in most societies today, it is not fair or effective to ascribe backward-looking responsibility to individuals taken separately or to the group of individual consumers or citizens in these societies.

Individual forward-looking responsibility

To abandon discussion about individual responsibility in the backward-looking sense does not necessarily entail abandoning all notions of individual responsibility for the environment. My claim is that although individuals should not be held responsible in the backward-looking sense if they did not have reasonable alternatives to act in environmentally friendly ways, it is nevertheless sometimes reasonable to ascribe forward-looking responsibility to individuals.

As mentioned in Chapter 2, Garvey suggests that just as "ought implies can", which is a core principle in ethics, "can implies ought" in some circumstances (Garvey, 2008). This is what I have conceptualised as forward-looking responsibility. The idea that you should contribute more if you have more resources or a greater ability to contribute to solving a problem is endorsed by many people, although not necessarily explicitly. As I argued in Chapter 2, I think there is a common intuition in contemporary societies that if an agent, whether individual or collective, has the capacity, power and resources to contribute to solving a social problem, then they have a responsibility to do so. Simply put, power and capacity entail responsibility. This kind of responsibility is forward-looking and fairly open-ended, and it is not necessary to state exactly when such a responsibility has been fulfilled (Goodin, 1986). Consequently, it leaves open exactly what the agent ought to do and a certain degree of flexibility and improvisability is required to resolve this question. In my view, this is the way in which individuals as consumers and citizens are responsible for environmental problems. Individuals should do what they can against the background of their particular situation. The more resources, power and capacity an individual has, the greater her ability to contribute to solving the problem, and the more reasonable it is to ascribe forward-looking responsibility to her. In this view, causal links are less central than they are in the backward-looking concept. Just as Bill Gates did not cause the social problems he contributes to solving by giving money to charity, the individuals who are best placed to contribute to solving environmental problems may not be the ones who contributed most causally to these problems. Their position to do more can be due to several factors, most likely involving financial resources but also education, information and leadership skills in terms of a talent for influencing others.

Collective responsibility for environmental problems

We now have the conceptual and normative tools to examine how collective agents like governments and corporations are responsible for environmental problems, and how their responsibility is related to individual responsibility. Governments and corporations are responsible because it is in their power to create reasonable alternatives for individuals.

They have it in their power to make it easier and less expensive for individuals to choose the environmentally friendly option, and they can provide information that is easily accessible and straightforward. In essence, they are responsible because they have the power to create opportunities for individuals to do what is right. Another way of phrasing it is to say that governments and corporations can make it easier for individuals to assume forward-looking responsibility. This can be done by making information accessible, by subsidising organic food while taxing non-organic food, by developing new products and by framing and presenting risks, benefits, products and services. The greater the extent to which these actors have done their share, the greater the extent of individual responsibility. The greater the extent to which institutional agents have taken care of their forward-looking responsibility, the greater the extent to which it is reasonable to ascribe both backward-looking and forward-looking responsibility to individuals when they do not choose the environmentally friendly option. This is the case for two reasons. First, the greater the availability and affordability of good options, the more reasonable it is to blame those individuals who still do not adjust their behaviour. Second, the greater the extent to which institutions such as governments have assumed their responsibilities, the larger the group of individuals who have enough capacity and resources to assume their forward-looking responsibility will be. For example, if the government has invested in an extensive public transport system, the degree of individual responsibility for choosing to drive a car instead of using public transport is higher than it would have been if the public transport system was underdeveloped and unreliable.

If a corporation can provide environmentally friendly products at a reasonable cost, it is their responsibility to do so. For example, restaurants should be able to provide customers with food that has been produced in environmentally friendly ways and not transported across the globe. Some critics will argue that personal responsibility will be eroded if we allocate too much responsibility to institutional actors. Do individuals not have to do anything by themselves? I think there are two ways to respond to such criticism. One way of conceptualising a forward-looking responsibility is through a virtue-ethical approach. Williams views responsibility as a virtue, which essentially represents a "readiness to respond to a plurality of normative demands" (Williams, 2008). This is a slightly different way of conceptualising the idea that individuals are too complex to assess morally merely on the basis of isolated actions. Instead, the focus should be on an individual's whole life and character as well as the way in which the character evolves and improves. We will explore responsibility as a virtue in Part II. Suffice it to say, at this point, that the focus in the environmental context should be on the challenges that individuals face when trying to respond to the plurality of, sometimes conflicting, demands. As the examples have illustrated, being a socio-economically underprivileged single parent may put one in a situation where it is extremely difficult to buy the most sustainable and healthy food for one's family. Similarly, it is not always easy to know that what is healthy, e.g. salmon, is the most sustainable option. Norms and expectations put the individual in constant dilemmas. Even when viewed from that angle, personal responsibility is still very important. However, what should be acknowledged is that an individual's character and the virtue of behaving responsibly can be affected by social systems, policies, access to information, and education.

Against this background, collective environmental responsibility can be stated as follows:

> Governments and corporations are responsible for creating systems and infrastructure to make it easier for individuals to respond to, or even to comply with, the emerging reasonable norm that we ought to act in environmentally friendly ways.

This demand for collective responsibility could also be described in terms of "nudging", i.e. creating choice architectures that facilitate good choices (Thaler and Sunstein, 2008).

A second way of responding to the critique that personal responsibility is eroded if too much responsibility is ascribed to governments and corporations is the following. Ascribing and distributing responsibility is a social practice that serves two purposes. First, it creates or establishes fairness. Second, it is a tool to establish an effective and efficient division of labour in order to solve societal problems. Williams (2008) talks about a moral division of labour in a discussion of responsibility as a virtue. According to Williams, the institutional fabric of liberal democratic societies creates a moral division of labour by delimiting spheres of responsibility. As we saw in previous chapters, the optimal distribution of responsibility is both fair and effective, although it is sometimes difficult to achieve both to the same extent. Sometimes, one of the two purposes is more important than the other and the two have to be weighed against each other in each case. There are areas of life and society where stating that something is the personal responsibility of individuals is fair and when fairness is the most important feature of the distribution of responsibility. There are other contexts, such as climate change, where the problems are too urgent and too vast to only focus on fairness and personal responsibility. The long-term goal should be to encourage individuals, as citizens and consumers, to be virtuous; for example, to encourage people to embrace green virtues (cf. Jamieson, 2007; Ladd, 1991). In order to achieve a sustainable society, we all need to think more about consuming less energy, for example by reducing air travel and by not eating food that has been transported across the globe. It would obviously be welcome, and necessary in the long run, if people in general would start to care about the environment more naturally and consistently. However, from the short-term perspective we must add that the distribution of responsibility should also be effective and efficient, that is, should contribute to solving the problem. That is why the greatest share of responsibility for environmental problems should be ascribed to the most powerful, resourceful and capable actors, i.e. governments and corporations, because they can create systems that make it easier and less costly for people to choose the environmentally friendly option than to choose the environmentally harmful option.

As argued by Shue, some duties should be assigned to institutions instead of individuals because it is likely to be more efficient. Institutions can enable the coordination and cooperation that are needed for those duties to be fulfilled. A second reason is that leaving the problem of climate change to individuals would demand too much of them as a duty to fulfil (Shue, 1988).

Thus, for reasons of efficiency as well as for reasons of fairness, institutions should be considered the main duty-bearers, as opposed to individuals. However, this does not mean that individuals are completely exempt. On the contrary, it is their duty to make sure that institutions are adequate to fulfil the duties in question (Shue, 1988).

Green argues that the responsibility of institutions is greater than the responsibility of individuals. He argues that while it is reasonable to hold to the restrictive version of responsibility – that is, to the notion of responsibility that traces behaviour to harm for individuals – a more comprehensive kind of responsibility should be assigned to institutions because they constitute a different kind of agent. Governments, for example, have more power and can alter mass behaviour, they are better at collecting and processing information about direct and indirect consequences of their actions and they can spread the cost through taxation. Essentially, institutional agents have more capacity; hence a greater share of responsibility is justified (Green, 2005). Similarly, Sinnott-Armstrong argues that whereas individuals do not have a moral obligation not to waste gas, governments have a moral obligation to fight global warming, primarily due to the scale of the problem.

Thus, due to the urgency and scale of environmental problems, it appears a good case can be made to include collective actors in discussions about how to distribute responsibility for such problems. Having said that, the fundamental unit in society is the individual citizen, and collective actors are created and upheld by individuals acting together. Hence, it also appears reasonable to expect individuals with capacity, resources and knowledge to create environmentally friendly institutions. In addition to voting, this also involves creating, supporting and joining organisations that work to improve the environment.[2] This, in turn, could lead to different individual activities, such as working to directly improve the environment, working to improve consumers' options and increase their access to information, working to raise people's awareness and working to change the perspectives of more traditional organisations (trade unions, culture and sports associations and so forth). As argued by

Sinnott-Armstrong, instead of just withdrawing from society and adjusting one's own lifestyle to create as little environmental damage as possible, it is even more important to be proactive and to work to change government policies and laws (Sinnott-Armstrong, 2005).

I have criticised the increasing tendency to hold individuals responsible for environmental problems in a backward-looking sense. I have argued that individuals are not appropriate targets of blame when acting in environmentally destructive ways unless they have reasonable alternatives. Today, many individuals lack such options or do not have the resources to do the environmentally friendly thing. This is the case in poor countries, but certainly also in some areas and among certain groups in rich countries. Here are just a few general or structural obstacles to individuals in modern societies that make it unreasonably difficult to act in environmentally friendly ways.

- The infrastructure in many societies encourages people to drive instead of using public transport or bicycles. Individuals need to drive to the supermarket, to work and to school. It is often easier and/or less expensive to drive than taking the train.
- It is assumed in many industries that people need to meet face to face, which leads to extensive business travel.
- Information about the origin and energy cost of producing certain consumer goods such as food is often inadequate.
- Government information is sometimes unclear, or even conflicting. Food products that are promoted for health reasons are sometimes discouraged from an environmental perspective. A case in point is the Swedish National Food Administration's quite complicated recommendations. People are encouraged to eat more fish, but to avoid fish with dioxins. In addition, the agency claims that it is beneficial for health to eat fish two to three times a week, but that it is better not to eat fish too often for environmental reasons. So, you should eat fish, and because wild fish can be toxic, you should eat farm-raised fish, but eating farm-raised fish is not good for the environment (Livsmedelsverket, 2017).
- Climate-smart food is often substantially more expensive than regular food.

These are just a few general structural problems; there are also individual differences that should be taken into account. An individual's options and the cost of acting in an environmentally friendly way should be seen as interplay between the individual and her surrounding socio-economic, political and cultural environment.

Although holding individuals responsible in the backward-looking sense is questionable, holding individuals responsible in a forward-looking sense is reasonable. Again, the different contexts and the different extent to which individuals have the capacity and resources to assume such responsibility should be taken into account. The most important conclusion is that governments and corporations have a great degree of forward-looking responsibility to create opportunities for individuals to behave responsibly and to act in environmentally friendly ways. Although acknowledging individual responsibility is beneficial, we should make sure that collective responsibility is not overlooked or ignored as a consequence.

Conclusion

As we have seen in previous chapters, there are good reasons to argue that responsibility ascriptions and distributions should be both (i) fair and (ii) effective. They should be fair for reasons of social cooperation and humanity. Cooperation is easier to achieve in a society where the norm is to hold someone responsible only when it is fair to do so, and such a society is arguably more humane. Of course, there is no simple answer to the question of when it is fair to hold someone responsible. However, it is common both in social practice and philosophical discussions to apply a number of conditions when ascribing responsibility. For instance, an agent should only be held responsible if she is a moral agent, meaning that she is a mentally sound adult, she contributed causally to the event, she knew what she was doing, she did it voluntarily and what she did was wrong according to some set of norms (Van de Poel *et al.*, 2012). Of course, there is disagreement about the content of these conditions as well as about how important each one of them is. The point is that we commonly use some set of conditions when ascribing responsibility, and this can be seen as a way to make sure that responsibility is ascribed and distributed fairly.

The aim of efficacy concerns the way in which ascriptions and distributions of responsibility contribute to solving societal problems. Whether it be public health, poverty, education or the environment, when discussing to whom we should ascribe responsibility and the question of how responsibility should be distributed between different actors (individuals, governments, corporations, teachers, parents and so forth) we do not merely care about what is fair, but also about who would be able to solve the problem at hand. To elaborate on this notion, my conclusion can be stated as follows. It is not fair to ascribe responsibility in the backward-looking sense, i.e. to blame individuals for environmentally destructive actions unless they have reasonable alternatives and resources to act in environmentally friendly ways. However, it is fair to ascribe forward-looking responsibility to individuals based on their capacity to contribute to solutions to environmental problems. Furthermore, a considerable share of forward-looking responsibility should be ascribed to governments and corporations because they can expand the group of capable – and therefore responsible – individuals. The urge to ascribe forward-looking responsibility to collective agents is motivated by the aim of efficacy in the distribution of responsibility. Simply put, if we ascribe responsibility to governments and corporations, we have a better chance of creating a society in which the opportunities to act in an environmentally friendly way increase.

Notes

1 This section is largely based on the article Nihlén Fahlquist, J. 2009. Moral responsibility for environmental problems – Individual or institutional? *Journal of Agricultural and Environmental Ethics* 22(2): 109–124.

2 I would like to thank the Editor-in-Chief for *Journal of Agricultural and Environmental Ethics* for commenting on this.

References

Aristotle 2000. *Nichomachean Ethics* (translated and edited by R. Crisp). Cambridge: Cambridge University Press.

Bovens, M. 1998. *The Quest for Responsibility: Accountability and Citizenship in Complex Organisations*. Cambridge: Cambridge University Press.

Cooper, D. 1968. Collective responsibility. *Philosophy*, 43: 258–268. (Reprinted in L. May and S. Hoffman (eds), *Collective Responsibility: Five Decades of Debate in Theoretical and Applied Ethics*. Lanham, MD: Rowman & Littlefield, 1991.)

Fischer, J.M. and Ravizza, M. 1993. Introduction. In J.M. Fischer and M. Ravizza (eds), *Perspectives on Moral Responsibility*. Ithaca, NY: Cornell University Press.

French, P. 1987. Collective and corporate responsibility. *Philosophical Review* 96(1): 117–119.

Garvey, J. 2008. *The Ethics of Climate Change: Right and Wrong in a Warming World*. London: Continuum.

Goodin, R.E. 1986. Responsibilities. *Philosophical Quarterly* 36: 50–56.

Green, M. 2005. Institutional responsibility for moral problems. In A. Kuper (ed.), *Global Responsibilities: Who Must Deliver on Human Rights?* New York: Routledge.

Jacobsen, E. and Dulsrud, A. 2007. Will consumers save the world? The framing of political consumerism. *Journal of Agricultural and Environmental Ethics* 20: 469–482.

Jamieson, D. 2007. When utilitarians should be virtue theorists. *Utilitas* 19(2): 160–183.

Ladd, J. 1991. Bhopal: an essay on moral responsibility and civic virtue. *Journal of Social Philosophy* 32(1): 73–91.

Lewis, H.D. 1948. Collective responsibility. *Philosophy*, 24: 3–18. (Reprinted in L. May and S. Hoffman (eds), *Collective Responsibility: Five Decades of Debate in Theoretical and Applied Ethics*. Lanham, MD: Rowman & Littlefield, 1991.)

Livsmedelsverket (National Food Agency, Sweden). 2017. www.livsmedelsverket.se/matvanor-halsa-miljo/kostrad-och-matvanor/rad-om-bra-mat-hitta-ditt-satt/fisk/ (accessed 13 Febuary 2017).

Micheletti, M. 2003. *Political Virtue and Shopping: Individuals, Consumerism, and Collective Action*. New York: Palgrave Macmillan.

Pfeiffer, R. 1988. The meaning and justification of collective moral responsibility. *Public Affairs Quarterly* 2(3): 70.

Shue, H. 1988. Mediating duties. *Ethics* 98(4): 687–704.

Sinnott-Armstrong, W. 2005. It's not my fault. In W. Sinnott-Armstrong and R.B. Howarth (eds), *Perspectives on Climate Change: Science, Economics, Politics, Ethics*. Amsterdam: Elsevier.

Strawson, P.F. 1962. Freedom and resentment. *Proceedings of the British Academy* 48: 1–25 (reprinted in J.M. Fischer and M. Ravizza (eds), *Perspectives on Moral Responsibility* (pp. 45–66). Ithaca, NY: Cornell University Press, 1993).

Thaler, R. and Sunstein, C. 2008. *Nudge: Improving Decisions about Health, Wealth, and Happiness*. New Haven, CT: Yale University Press.

Van de Poel, I., Nihlén Fahlquist, J., de Lima, T., Doorn, N., Royakkers, L. and Zwart, S. 2012. The problem of many hands: climate change as an example. *Science and Engineering Ethics* 18: 49–67.

Williams, G. 2008. Responsibility as a virtue. *Ethical Theory and Moral Practice* 11(4): 455–470.

Part II

Taking responsibility for risk

Introduction to Part II

Taking responsibility for risk

In Part I, I analysed three kinds of questions concerning the ascription and distribution of responsibility for risk. When analysing a certain distribution of responsibility, we can, and should, ask questions concerning (a) fairness and efficacy, whether it is a (b) backward-looking or forward-looking kind of responsibility, as well as considering to what extent something is an (c) individual or collective responsibility. The previous chapters attempt to clarify some of the conceptual nuances of the notion of moral responsibility as something to be distributed and its relevance to the area of risk.

One does not have to examine philosophical literature and discussions to notice the nuances and different meanings of the concept of responsibility. In everyday discussions and public debates, we do not merely talk about responsibility as backward-looking blame or forward-looking task distribution. The picture is not complete until another dimension of risk and responsibility is added, which will be explored in this part of the book. In some cases, responsibility refers to certain individual character traits, or even possibly those of a collective actor. For example, we select a physician who is particularly empathetic, competent and thorough, or a teacher that our child trusts deeply and who helps the students flourish. We talk about a funeral agency that helped us arrange a beautiful funeral for a loved one. Responsibility is not just about blame, tasks or obligation, but also refers to a character trait, a group of character traits, or a way of doing things in a certain, excellent and admirable, way. We talk about *being* a responsible person or organisation, and about *taking* responsibility for things. We refer to *how* "responsible" people tend to be, how some people act "responsibly" and how others act irresponsibly. I will argue that these notions are most comprehensible in the context of virtue ethics.

Recall the discussion on individual versus collective responsibility for health and the environment from Chapter 3. I discussed how the primary collective responsibility is to create a good system – a choice architect to use Thaler and Sunstein's (2008) terms – thereby facilitating good, environmentally friendly and healthy choices. I argued that even though the best way to promote environmental responsibility would probably be to encourage people to develop and cultivate "green virtues" (Jamieson

2007), that this would be a rather lengthy project. The long duration of this project necessitates political action and collective forward-looking responsibility. In Part II, I will discuss this long-term challenge of cultivating a sense of responsibility, of promoting "responsibility-taking". If it would be a good idea to encourage for example green virtues, what does this mean for the individual? What does it mean to develop a sense of responsibility, conceived as a virtue or character trait?

According to Schmidtz, the crucial distinction when we talk about social problems and challenges is not the one between individual and collective responsibility, but the one between *internalised* and *externalised* responsibility. Responsibility is externalised, he argues, when people do not take responsibility for "the mess" they cause and find themselves in. In contrast, responsibility is internalised when people do take responsibility for their future and their own welfare (Schmidtz and Goodin, 1998). According to Schmidtz, externalised and internalised responsibility can be either individual or collective. The crucial aspect is whether one sees one's welfare as one's own or someone else's problem. We could replace the value of welfare with health, safety, or sustainability.

In response to this argument, Goodin asks what we should do about those people who do not assume this kind of responsibility. Even if this kind of internalised responsibility is a virtue and an ideal, we need to decide what to do when people do not live up to that ideal, he argues (Schmidtz and Goodin, 1998). Whereas Goodin prefers the notion of responsibility as conceptualised in Part I, that is, responsibility as task or "oughts" to be distributed, Schmidtz leans towards the virtue-ethical notion of responsibility that is explored in this part of the book. I agree with Schmidtz about the value of internalised responsibility and I am convinced that both notions are necessary in discussions concerning risk.

Thus, in this part of the book, I focus on the notion of actively taking responsibility for technological, environmental or health risks, and this notion is conceptualised within the framework of virtue ethics. First, I explore what responsibility as a virtue could mean. Second, I discuss how collective endeavours, such as engineering and research, can benefit from the ideal of internalised responsibility, that is, the cultivation and encouragement of responsibility as a virtue. The cultivation of responsibility as a virtue is needed in order to avoid what has been called the "problem of many hands", i.e. the ethically and practically troublesome situation in which there is harm for which no one is responsible. The "problem of many hands" is prevalent in many areas in society. When engineers, for example, take part in innovation and design processes that create tremendous opportunities but also risk, I will argue that we should expect them to take responsibility in the virtue-ethical sense.

Third, I analyse what responsibility as a virtue means for risk communication. If a government takes on all responsibility for issues such as public health or climate change, this is likely to affect people's *sense* of responsibility. If people start to trust the government to take care of all health and environmental problems, people may think less and less about what their responsibility is or how they can contribute. On the other hand, if people are told it is up to them whether they develop incurable diseases and whether climate change is ever managed properly, people may become apathetic or resigned. It appears a balance needs to be struck between personal responsibility-taking and institutional and systemic change.

Finally, I discuss how the adult world should encourage a sense of responsibility in children with regard to risk. When speaking of the need to cultivate and encourage "green virtues" or a sense of responsibility for one's own health, it becomes clear that this activity starts in early childhood. The long-term project of encouraging citizens of modern societies to develop responsibility as a virtue surely requires that children become moral agents and citizens. This is especially important given the increasing problems related to issues such as climate change and public health problems such as obesity. There is a balance to be struck between, on the one hand, encouraging children to take responsibility for their own health and to start thinking about their contribution to climate change, and, on the other hand, not blaming children for a situation that previous generations created. Fairness as well as efficacy are important in this context.

The inter-relationships between the different aspects of risk and responsibility are becoming clear. To trust personal responsibility-taking may be ineffective. In this way, questions having to do with taking responsibility are related to questions of individual versus collective responsibility as well as to the balance between fairness and efficacy. However, when it comes to the distinction of backward-looking and forward-looking responsibility, the ethical focus of "responsibility-taking" is mostly about forward-looking responsibility. It is about what an agent does in order to cultivate the virtue of responsibility-taking and what she does when acting responsibly.

Being a responsible person – insights from virtue ethics[1]

As I hope I have made clear in previous chapters, to "be responsible" can refer to "seeing to it that X", i.e. fulfilling one's obligations or tasks, or to a certain kind of person with a specific set of character traits, attitudes and emotions. Both are essential, and people who are "responsible" in the virtue-ethical sense are more likely to take on and fulfil moral obligations and tasks. However, the two meanings are not identical, and a closer look at the notion of being a responsible person is needed to grasp how it adds to what has been said so far.

When considered in the virtue-ethical sense, a person is responsible if she has certain character traits, attitudes and emotions and behaves in a certain way over time. One major difference between responsibility conceived as blameworthiness and responsibility conceived as virtue is that, whereas the former is backward-looking, the latter is primarily forward-looking. A difference between responsibility conceived merely as obligation and responsibility as virtue is that, whereas the former is more definable for a specific task or set of tasks, the latter is more open-ended and voluntarily assumed. Neither responsibility as blameworthiness nor responsibility as obligation or task pays attention to the agent herself or her character; instead they are attached to individual and usually isolated actions or states-of-affairs. For example, who caused the traffic accident? Alternatively, who should improve road traffic safety? To be a responsible person is to be the kind of person who acts in certain ways over time, perhaps even a lifetime.

So, what character traits does responsibility as virtue involve? The suggestion here is that responsibility as virtue should be seen as primarily requiring care, moral imagination and practical wisdom. This conception brings emotions to the fore of responsibility. Unless one cares or empathises with potential victims of the unintended as well as the intended consequences of one's actions, behaviour and activities, one does not act responsibly in the virtue sense. In order to be a responsible person in this sense, one must *feel* responsible, and in order to feel responsible, one needs to *care*. Or so I will be argue. Let us start by analysing the basic intuition that gives rise to this notion.

Responsibility as a virtue – a first sketch

Let us start with an illuminating example having to do with events that entail technological and environmental risk. The Deepwater Horizon oil spill that occurred in April 2010 was not the first time BP's activities had caused harm in terms of lives lost, health and environmental problems. They had evidently previously neglected to adequately upgrade the safety systems at their Texas City Refinery, causing fatalities on several occasions. The most well-known accident prior to the Deepwater Horizon oil spill occurred in March 2005, when one of their largest refineries in Texas City exploded. Management had allegedly not paid enough attention to engineering problems, and maintenance and safety had not been upgraded according to best practices for reasons of cost (Hopkins, 2011). BP ignored safety and environmental regulations and had not implemented necessary improvements, their safety culture and safety practices were flawed and there were poor management decisions (cf. Baker *et al.*, 2007; Bakolas and Saleh, 2011; Steffy, 2011). It appears as though BP's management may not have acknowledged all the risks and therefore acted in ways that are potentially blameworthy. It is clear that most people feel a need to blame someone in cases like this one, where BP's management was judged to have done things wrong, for example because they had information on how to improve safety but chose not to do it. In such a case, conditions of responsibility as backward-looking blameworthiness are fulfilled, and assigning blame could be considered appropriate. However, when backward-looking responsibility becomes relevant, it is usually because someone neglected to take forward-looking responsibility, and this is arguably what led to the unwanted situation. That some people did not take forward-looking responsibility can mean that (1) they violated their obligations to do x, y and z (responsibility as moral obligation or task), or (2) that they did not act in a responsible manner (responsibility as virtue). Whereas obligations and tasks are relatively specific, being a responsible person or acting responsibly can require doing something beyond what the obligations prescribe. In the following discussion, I will explore responsibility conceived in this way.

Most people probably find it reasonable that when someone takes part in an activity that creates risks, for example oil production, she has, or should have, a higher degree of responsibility. This view, I would argue, is justified. Changing the status quo by creating opportunities and risks entails moral responsibility. Arguably, we expect that certain professions whose activities affect people's lives and health are responsible to a greater degree than people who do not have that power to change people's lives in a good or bad way. This partly means that those whose actions cause development and risks should be praised when things go well and blamed when things go wrong. When we talk about responsibility in this way, we are referring to backward-looking responsibility because we are pointing at what was done and the consequences of it.

However, intuitively, responsibility for risk-generating activities also means taking responsibility during the very process in which risks are generated. In the case of BP, it would be reasonable to expect employees to step up and do something to stop the neglect of safety and environmental concerns. Some employees may have done so, but unfortunately that did not change the outcome. It is reasonable to think that people working in the research and development of risky technologies are *responsible* people, and that if they see something that could potentially be dangerous, they do something about it, in other words, that they make a serious effort, deliberate about potential risks and uncertainties, and do not act recklessly or are not too risk-prone or too "adventurous". It is reasonable to hope and expect that people working in risk-generating activities are empathetic and have the ability to think about the potential victims of materialised risks. However, it is not common to think of engineers as empathetic. It is important to note that the image of an engineer does not necessarily have anything to do with the personality of people who are engineers. Engineers are probably as empathetic as non-engineers. However, engineers are commonly seen as making decisions in a rational and quantifiable way, meaning that emotions have little to do with engineering and should perhaps be left out completely. We do not generally expect engineers to be emotional or caring, at least not in their professional roles.

However, as Roeser argues, there are good reasons to think that engineers should cultivate their moral sensitivity by learning to use their emotions. This should be done because engineering design is not morally neutral but involves values (Roeser, 2012). This ability arguably requires a certain emotional maturity and empathetic attitude, that is, being a certain kind of person. It is reasonable to think that responsible people, in this case engineers, both seek to avoid unnecessary risks to people's health, lives and the environment and that they do everything in their power to reduce the negative impact of their activities. Thus, partaking in risk-generating activities entails a forward-looking responsibility, a sense of responsibility.

The aim in the remainder of this introduction to the final chapters is to explicate the notion of responsibility as a virtue and to sketch what responsible people could be like. To do that, I will start by having a brief look at virtue ethics. As stated above, the reason for turning to virtue ethics is that this is the ethical theory that is concerned with character traits, emotions, attitudes and people's behaviour over time as opposed to utilitarianism and deontological ethics, which are more concerned with the moral quality of isolated actions.

Virtue ethics

Virtue ethics is usually seen as the third major ethical theory, the other two being deontological ethics and consequentialism. Deontological ethics is based on the notions of duties and rights and, to state its premise very simply, judges an action to be right if it is in accordance with universal duties and rights, and wrong if it violates a duty. Consequentialism, again to state its premise very simply, evaluates actions according to their outcomes. An action is right if it contributes to happiness or utility and wrong if it produces the opposite, that is, unhappiness or disutility. Whereas these ethical theories emphasise actions, virtue ethics is concerned with the person and her character.

The word virtue means "excellence", "capacity" or "ability" and being virtuous means being able to or having the power to achieve something. Virtue ethicists are concerned with traits of character that are admirable, however, the lack of which would not necessarily lead to blaming or accusations of acting immorally. A moral agent is thought to be motivated to the extent that she *wants* to be a morally good person and because she recognises certain values, such as honesty. Virtue ethicists attempt to find answers to questions of what an agent should do by considering the agent's character and the morally relevant features of the situation (van Hooft, 2006). Plato and Aristotle explored the virtues during Antiquity, and virtues like courage, temperance and justice were important during that time. Even though we do not really talk about virtues in modern society, we do talk about desirable and undesirable character traits.

What virtue ethicists have in common is the view that possessing virtues is the basis of being good as a human being. According to Swanton (2007), "a virtue is a disposition to respond to or acknowledge items in its field or fields in an excellent or good enough way". Excellence includes not only actions, but also emotions and motivations. Swanton defends a pluralist account of virtues, which entails that virtues make the possessor excellent in responding to the "demands of the world", including demands of the self. Virtue ethics has been criticised for lack of action-guidance. This has led some philosophers to work on the action-guiding potential of virtue ethics (e.g. Hursthouse, 1999). Accordingly, Oakley and Cocking (2001) argue that the best way to understand the virtue ethics criterion for right action is in terms of "regulative ideals". An agent who has a regulative ideal has internalised a specific conception of correctness or excellence in a way that enables her to adjust her behaviour and motivation in accordance with that standard.

If a virtue is a disposition to respond in a certain way to the demands of the world, then it is concerned with people's behaviour over time. Against this background, what is required of a person to be called "responsible"? What would *responsibility* as virtue entail? In everyday situations, we talk about responsible and irresponsible people, leaders, parents, teachers, politicians and so forth. According to van Hooft (2006), who briefly discusses the notion, responsibility as virtue is about a "willingness to take responsibility" and about "being responsible", that is, being a responsible person. It is not the same as a willingness to accept blame for things an agent has done wrong, although that may be one part. The emotions involved in taking responsibility are personal involvement, commitment, and "not leaving it to others", a feeling that it is "up to me" and a "willingness to make sacrifices in order to get involved". Of course, defining responsibility as a "willingness to take responsibility" is circular and does not help us further in our exploration of the concept. However, to be fair, van Hooft is not trying to define responsibility as a virtue, but rather to indicate the relevance of the concept. The notion of "not leaving it to others" does help somewhat. A responsible person tends to take important matters seriously enough to not leave them and wait for someone else to do it. She is not a "free rider", if you will.

Williams (2008) has discussed the relevance of responsibility as a virtue in modern society. He argues that responsibility as virtue represents a "readiness to respond to a plurality of normative demands". The key to understanding responsibility in this sense is the notion of plurality. In a typical case of plurality, responsibility is required for more than one important demand, and the responsible agent needs to use her judgement to balance these demands in order to form an adequate response. Most of us fulfil various roles and expectations, as employees, parents, friends, citizens and colleagues at the same time, and being responsible is about juggling the different demands raised by these different roles in a sensible way and responding in appropriate ways (Williams, 2008). As the complexity of society increases, this approach to ethics becomes even more applicable.

Discussing one of the greatest industrial disasters in history, the Bhopal accident in 1984, Ladd (1991) conceived of responsibility as connected to "civic virtue":

> A virtuous citizen, and that should include everybody, should have a concern for the common good and for the long-range welfare of other people in the society, even where this concern demands individual sacrifices of one sort or another or simply giving less priority to one's own private interests and to one's advancement on the escalator to worldly success.

There are some theories of responsibility that are related to the idea of responsibility as virtue without explicitly referring to virtue ethics. First, Young (2006) talks about responsibility due to "social connections". Second, Fredriksen (2005) discusses responsibility due to "social involvement". What is common in these approaches is that responsibility as virtue is (1) forward-looking, (2) focused on the person and her relations to other people and the world as opposed to individual actions, (3) requires that the person see herself as part of a greater context within which she acts and (4) requires that she act in a certain way, however acting responsibly is defined, over time. A virtuous-responsible person *feels* responsible for more than she directly causes. This does not necessarily mean that she fears blame or would be blamed in case something happened, nor that she needs to feel responsible for others to the extent of self-effacement. It does mean, I would argue, that she cares about others and that she has adequate moral imagination and practical wisdom to turn her attitude into responsible behaviour. I will discuss these character traits below.

Care, moral imagination and practical wisdom

I propose that an agent acts responsibly in the sense of a virtue if: (i) the agent *cares* about other people and the way the activities in which she partakes potentially affect other people (and the environment), (ii) the agent has the emotional ability to *morally imagine* what those effects could be like and what risks might be involved in those activities and (iii) the agent has the cognitive ability to transform these concerns into practice and actions, i.e. she has *practical wisdom*.

When someone partakes in an activity that might risk other people's health and well-being (whether or not they are aware of the risk) it is reasonable to ask this person to have the intellectual and emotional capacity to imagine potential victims and the risks that may be generated and that the person cares about her fellow human beings. In addition, the agent should be able to transform those emotions and attitudes into actions that go beyond good intentions. If the agent cares, but can't act wisely on that emotion, care becomes merely an attitude or intention, while responsible behaviour is about actions as well as intentions and attitudes.

Taking BP during the years between 2005 and 2010 (in order to include the Texas refinery explosion and the Deepwater Horizon disaster but without saying anything about what BP is like now or in the future) as an illustrative example, what character traits would have been desirable on behalf of the managers and employees to possess? How would we have wanted them to respond to what was going on at their company, the risks involved and the demands of the world? It would have been reasonable to expect that someone who is involved in a business whose consequences are potentially disastrous *cares* about how their actions might affect the world, that they have the ability to *imagine* the risks and how the materialised risks would affect the world, and that they have the abilities and capacities necessary to transform these into actions, in other words, that they possess *practical wisdom*. Let us now discuss these intellectual-emotional character traits in turn.

Care

Care, which is the central notion in the normative theory called the ethics of care, can be seen as a virtue, a relation, or a practice (cf. Held 2006; Noddings 2002; Slote, 2001; Slote, 2007). It seems reasonable to say that care involves all these aspects. It includes emotions and attitudes associated with a virtue, but it is also important to take into account whether the activity labelled "care" has the intended effects and it should involve or lead to some kind of activities. This means that it involves an attitude or concern for other human beings, an acknowledgement that how that concern is received matters, and a method for transforming the concern into action. The ethics of care emphasises the central role of relationships and, initially, it arose as a critique against the allegedly "male", conventional and universalist rule- and principle-based moral philosophy. Gilligan's book *In a Different Voice: Psychological Theory and Women's Development* (Gilligan, 1990), first published in 1982, argued that women have a different way of thinking about moral matters, and that this "female morality" should be acknowledged as an important supplement to traditional "male" ethical theories. On the basis of her psychological research, she argued that women evidently emphasise care and connection to others instead of the male focus on autonomy, universal rights and justice. Inspired by Gilligan's research, philosophers began to theorise what these contrasting, or "female", ethical notions could and should mean.

In the beginning, the philosophical ethics of care focused on the practice of caring for those who are close to us and the kind of caring work that has traditionally been carried out by women. However, the contemporary ethics of care is generally applied to wider contexts and shows how care can be seen as the central notion in ethics. It is currently not viewed primarily as a "female" kind of theory, but as potentially universal and all encompassing, relevant not only to family life, but also to global issues such as social and foreign policy.

Noddings (2002), who initially argued that the intimate focus on caring – for example for our family – was important and that ethics of care was not applicable to social and political contexts, has changed her perspective over the years. She now argues that ethical caring – that is, the caring-*about* that characterises caring for people outside of one's intimate relations – should be seen as the very foundation of social policy. Caring-for is, she argues, the kind of caring that takes place in "ideal homes" and caring-about is the kind of caring that we engage in with distant others, i.e. other people whom we do not know, yet care about. Caring-about is dependent on caring-for, since we learn to care-about through our experience of being cared-for, and therefore, caring-about may be seen as the link between caring and justice:

> caring-about (or, perhaps a sense of justice) must be seen as instrumental in establishing the conditions under which caring-for can flourish. Although the preferred form of caring is cared-for, caring-about can help in establishing, maintaining, and enhancing it. Those who care about others in the justice sense must keep in mind that the objective is to ensure that caring actually occurs. Caring-about is empty if it does not culminate in caring relations.
>
> (Noddings, 2002, pp. 23–24)

To introduce the notion of care, which has traditionally been associated with the private sphere occupied by women, into the public sphere challenges the bureaucratic ideas that modern liberal democratic states are based on, including justice as an objective value promoted to oppose the subjective, and hence potentially arbitrary, slant of care. However, justice and care do not have to be seen as mutually exclusive. Slote (2007) attempts to reconcile care and justice, arguing that "the social virtue of justice" is a "function of individual virtue(s), i.e., of the virtuousness of the individuals who constitute a given society" (Slote, 1998, p. 188). Nevertheless, there is a difference between the traditional ethical theories attempting to universalise principles and rules on the one hand and the emotion-based notion that caring about the particular other (instead of a generalised individual) is the foundation of ethical thinking. Noddings describes it as follows:

> We would want children to avoid the unnecessary infliction of pain because they have learned to share the pain of others and be distressed by it, not because they fear retributive pain to themselves ... A sense of

> fairness will also be cultivated in good homes, but it is not fairness that should prevent the infliction of pain.
>
> (Noddings, 2002, p. 145)

Although Noddings talks about children and upbringing here, it is relevant to adults and society as well. Care is then closely related to empathy. Slote argues that care, understood as empathy, can work for the whole of ethics, and that one can have a caring attitude "towards (groups of) people one is never going to be personally acquainted with, inasmuch as one is genuinely, altruistically concerned or worried about what happens to them" (cf. Held, 2006). This is important for our purpose of situating care in the context of responsibility as a virtue. If care was only relevant in relation to our close family and friends, it would not be appropriate to see it as essential to responsibility because, clearly, we talk about and *should* talk about responsibility in relation to distant others or others who are unknown to us. The crucial thing is that we can imagine those others and care what happens to them even if we never get to know them personally. This does not mean that our caring emotions and efforts are equally strong for distant others as they are for our family and close friends. There is a difference in strength between humanitarian concern for others and caring for our loved ones. This difference is natural and morally permitted (Slote, 2007).

To be a responsible person is to feel responsible and feeling responsible requires caring about others' well-being and about how one's actions and the activities one takes part in affect fellow human beings as particular others. Little (1995) argues that caring is not only important but is a necessary condition for seeing the moral landscape. The emotional component is what distinguishes acting from duty or a sense of duty on the one hand and acting responsibly on the other.

To view responsibility as a virtue involving care is highly relevant to the context of technological risk. I will explore this in the following chapter. Suffice it to say, at this point, that in the context of the role of emotions in technological risk, important work has been done by, for example, Roeser (2006, 2012), who argues that engineers who work with risky technologies should use their emotions in order to judge the moral acceptability of risks. As famously argued by Nussbaum, emotions are cognitive as well as affective (Nussbaum, 2001; Roeser, 2011). Furthermore, emotions are judgements of value (Nussbaum, 2001).

If we want engineers to be responsible, and if we argue that one essential part of this is that they care about potential victims, then naturally they need to use their emotions. It is simply not possible to care and empathise without using emotions. Virtues have been described as "good habits of action *and* emotion" (Fritz Cates, 2003, p. 326). The importance of emotions to moral psychology has been discussed thoroughly in recent decades. For example, there is a debate about whether psychopaths have "cognitive empathy", but lack "emotional empathy" (e.g. Blair, 2005). The concepts of empathy, care, compassion and sympathy are under debate in philosophy too. Whereas Slote favours empathy, Nussbaum (2001), for example, prefers compassion.

I think, for this present purpose, it is not necessary to determine which concept is the correct one. It should merely be noted that care, involving empathy and compassion, is both affective and cognitive, that it makes us recognise the needs and concerns of other human beings, that it facilitates ethical analysis and that it is required for acting responsibly.

What is necessary is that one cares about the ways in which one's own actions and activities affect other people's well-being. If I am a responsible-virtuous person, I am able to put myself in other people's shoes, and it matters to me how they feel and how they are affected by the activities I take part in. I can act from a sense of duty by using wisdom and by applying principles and rules: reason, as it were. However, if I act responsibly in the sense presented here, I have to take many different normative demands into consideration and I need to use my emotions in order to know how to respond to a situation. In my view, emotions are crucial to this kind of care and, hence, crucial to responsibility as a virtue.

Moral imagination

As mentioned above, we can care not only for people we know well, but for people we will never know. What makes this possible is our ability to empathise and sympathise with other people's well-being and misery, happiness and sorrow. The emotional ability to care about the potential victims of risky activities also requires moral imagination, that is, the ability "to imagine how your actions influence others, being able to empathically understand others, and being able to envisage alternative courses of action if necessary" (Coeckelberg, 2006, p. 253). As argued by Coeckelberg (2006, p. 237), "[t]he enhancement of moral imagination can help engineers to discern the moral relevance of design problems, to create new design options, and to envisage the possible outcomes of their designs". Moral imagination has a significant role in the exploration of possibilities and in the projection of future scenarios (Coeckelberg and Mesman, 2007), which also entails the consideration of the concerns of the wider society (Davis, 1989). This is especially important in engineering and innovation contexts since those involve a lot of uncertainties. The problem with new technology is, of course, that we cannot always have a reliable estimate concerning risks, since some older technologies may be similar in some respects, but different in others. There is a lack of information or knowledge about the consequences of what one is deciding and doing.

Coeckelberg and Mesman (2007) discuss another context in which risks and uncertainties are a salient feature of the professional environment, namely, the neonatal ward at a hospital, and they argue that imagination is needed in order to fill the gap left by a lack of information, in their example whether the baby would survive, die or survive with a severe handicap. What these different contexts have in common is that professionals working within engineering and health care must confront uncertainties and sometimes must improvise, estimate and manage risks with potential consequences for people's health and well-being. People who are able to imagine risks will ideally be somewhat more cautious in their activities. According to Werhane (2002), moral imagination can help someone to disengage from a situation and to think creatively within the constraints of what is morally possible.

Werhane (2002) argues that moral imagination is not only beneficial for individuals, but that it is also applicable to organisations and systems. She illustrates the need for moral imagination with Eskom, a South African power company, which was an all-white managed company. She describes Eskom as a successful company during the apartheid system providing services only to the white communities. Even before apartheid ended, Eskom started to question and re-evaluate its own practices and came to the conclusion that they should provide electricity to the whole nation and train non-whites for management positions. Said and done, Eskom found ways to provide electricity to the poor non-white South African people and is today a very successful company. Eskom appears to have become morally imaginative, and it seems reasonable to think they also cared about people enough to be willing to make a change. They acted responsibly. In addition to this, they were also able to transform this concern into actions that changed the company and potentially contributed to moral progress in society as a whole. This brings us to the last component in responsibility as a virtue.

Practical wisdom

If an agent cares and has the ability to imagine how the activities in which she partakes could affect people and the environment, does that mean that she is a responsible person or a responsible member of an organisation or network? Care and moral imagination are the emotional-attitudinal requirements of responsibility as virtue. However, they need to be combined with a way to transform them into practice. Care needs to be focused in order to be effective and useful. Noddings (2002) points out that the relational aspect of care is important to always keep in mind, that is, how the recipient of care experiences the caring efforts. There is always the risk of care being too superficial or merely rhetorical ("I care *so* much for people in poor countries!") or too intrusive and paternalistic ("I care so much about my teenage children that I never allow them to go outside the house without me"). To be responsible, one needs to care in a relatively focused and effective way. This is why practical wisdom is also needed to complete the picture of what responsibility as a virtue requires.

Care is essential, but a responsible agent also needs judgement in order to prioritise her caring efforts. Judgment provides a compass and shows how to prioritise caring efforts and how to balance different concerns. Aristotle discussed phronesis as one of the intellectual virtues. Phronesis can be described as "administrative ability" (Pakaluk, 2005). Practical wisdom is less an ability to apply principles correctly than an ability to *see* situations correctly (Aristotle, 2000, pp. xxiv-xxvi). The person who possesses practical wisdom knows how to behave in particular circumstances, and this can never be reduced to knowledge of general truths. Phronesis involves deliberation about values and an orientation towards actions based on judgement and experience (Flyvbjerg, 2001).

Although virtue ethics is to a great extent focused on agents and the character of agents, actions are important as well. It makes no sense to talk about a responsible agent if this character trait never reveals itself or materialises in actions and behaviour. Phronesis in the context of responsible behaviour is what makes a person able to respond to the plurality of normative demands (cf. Williams, 2008) of the particular context within which she is situated. This skill is what makes it possible for the agent to prioritise and to decide what is more or less relevant.

When discussing responsibility as virtue, the question arises of whether or to what extent we are responsible for our character traits, i.e. the way we are. Philosopher Audi discusses whether we are responsible for our character and character traits and he distinguishes between generative responsibility, retentional responsibility and prospective responsibility to separate questions of producing a character trait, retaining a character trait and adopting a character trait. An agent can be excused from generative responsibility because she, for instance, did not produce the character trait. However, since she is now able to change it, is reasonable to ascribe retentional responsibility to her. So, even if she could not help becoming a certain way, the question of what she should do now and in the future remains open (Audi, 1991). In this way, she may be excused from backward-looking responsibility, but be ascribed forward-looking responsibility for that character trait.

The question of the extent to which we are personally responsible for changing the way we are despite, or perhaps because of, how we were wired becomes especially relevant in the context of epigenetics. The field of epigenetics studies how genes can be affected by the external environment, and although in its infancy as scientific discipline, it indicates that it could be the case that how we choose to live our lives may affect not only our own genes, but also the genes of our offspring. In this sense, epigenetics "blurs the conceptual boundaries between inheritance and development" (Beck and Niewöhner, 2006, p. 222). Consequently, it potentially puts "heavy demands on individual responsibility in terms of risk avoiding behaviour and contribution to costs of health care" (Hedlund, 2012). Interestingly, being responsible in the virtue sense of the concept may require not only caring about others, but also caring about oneself.

Questions of personal responsibility for lifestyle and developing our character in a health-promoting manner are likely to become increasingly important. This is particularly important in relation to children as moral agents in development. I will explore that context in the last chapter. Before that, I will discuss in what ways responsibility as a virtue is relevant in the context of technological risk and risk communication.

Note

1 The remainder of the introduction was first published as Nihlén Fahlquist, J. 2015. Responsibility as a virtue and the problem of many hands. In I. van de Poel, L. Royakkers and S. Zwart (eds), *Moral Responsibility and the Problem of Many Hands* (Chapter 6). London: Routledge.

References

Aristotle 2000. *Nichomachean Ethics* (translated and edited by R. Crisp). Cambridge: Cambridge University Press.

Audi, R. 1991. Responsible action and virtuous character. *Ethics* 101(2): 304–321.

Baker, J.A. III., Bowman, F., Erwin, G., Gorton, S., Hendershot, D.C., Leveson, N.G., Priest, S., Rosenthal, I., Tebo, P.V., Wiegmann, D.A. and Wilson, L.D. 2007. *The Report of the BP U.S. Refineries Independent Safety Review Panel.* Washington, DC.

Bakolas, E. and Saleh, J.H. 2011. Augmenting defense-in-depth with the concepts of observability and diagnosability from control theory and discrete event systems. *Reliability and System Safety*, 96: 184–193.

Beck, S. and Niewöhner, J. 2006. Somatographic investigations across levels of complexity. *BioSocieties* 1: 219–227.

Blair, R.J.R. 2005. Responding to the emotions of others: dissociating forms of empathy through the study of typical and psychiatric populations. *Consciousness and Cognition* 14: 698–718.

Coeckelberg, M. 2006. Regulation or responsibility? Autonomy, moral imagination, and engineering. *Science, Technology, & Human Values* 31(3): 237–260.

Coeckelberg, M. and Mesman, J. 2007. With hope and imagination: imaginative moral decision-making in neonatal intensive care units. *Ethical Theory and Moral Practice* 10(1): 3–21.

Davis, M. 1989. Explaining wrongdoing. *Journal of Social Philosophy* 20 (Spring/Fall): 74–90.

Flyvbjerg, B. 2001. *Making Social Science Matter: Why Social Inquiry Fails and How It Can Succeed Again.* Cambridge: Cambridge University Press.

Fredriksen, S. 2005. Luck, risk and blame. *Journal of Medicine and Philosophy* 30: 535–553.

Fritz Cates, D. 2003. Review: conceiving emotions: Martha Nussbaum's *Upheavals of Thought. The Journal of Religious Ethics* 31(2): 325–341.

Gilligan, C. 1990. *In a Different Voice: Psychological Theory and Women's Development.* Cambridge, MA: Harvard University Press.

Hedlund, M. 2012. Epigenetic responsibility. *Medicine Studies* 3: 171–183. https://doi.org/10.1007/s12376-011-0072-6

Held, V. 2006. *The Ethics of Care: Personal, Political and Global*. Cambridge: Cambridge University Press.

Hopkins, A. 2011. Risk-management and rule-compliance: decision-making in hazardous industries. *Safety Science* 49: 110–115.

Hursthouse, R. 1999. *On Virtue Ethics*. Oxford: Oxford University Press

Jamieson, D. 2007. When utilitarians should be virtue theorists. *Utilitas* 19(2): 160–183.

Ladd, J. 1991. Bhopal: an essay on moral responsibility and civic virtue. *Journal of Social Philosophy* 32: 73–91.

Little, M.O. 1995. Seeing and caring: the role of affect in feminist moral epistemology. *Hypatia* 10: 117–137.

Noddings, N. 2002. *Starting at Home: Caring and Social Policy*. Berkeley, CA: University of California Press.

Nussbaum, M.C. 2001. *Upheavals of Thought: The Intelligence of Emotions*. Cambridge: Cambridge University Press.

Oakley, J. and Cocking, D. 2001. *Virtue Ethics and Professional Roles*. Cambridge: Cambridge University Press.

Pakaluk, M. 2005. *Aristotle's "Nicomachean Ethics": An Introduction*. Cambridge: Cambridge University Press

Roeser, S. 2006. The role of emotions in judging the moral acceptability of risk. *Safety Science* 44: 689–700.

Roeser, S. 2011. Nuclear energy, risk, and emotions. *Philosophy & Technology* 24, 197–201.

Roeser, S. 2012. Emotional engineers: toward morally responsible design. *Science and Engineering Ethics* 18(1): 103–115.

Schmidtz, D. and Goodin, R.E. 1998. *Social Welfare and Individual Responsibility*. Cambridge: Cambridge University Press.

Slote, M. 1998. The justice of caring. In: E.F. Paul, F.D. Miller and J. Paul (Eds), *Virtue and Vice*. New York: CUP.

Slote, M. 2001. *Morals from Motives*. Oxford: Oxford University Press. www.heraldtribune.com/article/20110208/columnist/102081017

Slote, M. 2007. *The Ethics of Care and Empathy*. Abingdon: Routledge.

Steffy, L.C. 2011. Drowning in Oil: BP and the Reckless Pursuit of Profit. New York: McGraw Hill.

Swanton, C. 2007. Virtue ethics, role ethics, and business ethics. In R.L. Walker and P.J. Ivanhoe (eds), *Working Virtue: Virtue Ethics and Contemporary Moral Problems* (p. 209). Oxford: Oxford University Press.

Thaler, R. and Sunstein, C. 2008. *Nudge: Improving Decisions about Health, Wealth, and Happiness*. New Haven, CT: Yale University Press.

van Hooft, S. 2006. *Understanding virtue ethics*. Chesham: Acumen.

Werhane, P.H. 2002. Moral imagination and systems thinking. *Journal of Business Ethics*, 38: 33–42.

Williams, G. 2008. Responsibility as a virtue. *Ethical Theory and Moral Practice* 11(4): 455–470.

Young, I.M. 2006. Responsibility and global justice: a social connection model. *Social Philosophy and Policy* 23(1): 102–130.

4 Taking responsibility for technological risks

Introduction

Technological design and innovation can reduce risks but can also create new ones. Whether it will primarily reduce or create risks may not be clear from the beginning. For example, consider the automobile, which was created to make transportation easier and more efficient, thereby reducing the economic risks of having to rely on horses or trains. At the same time, the new infrastructure that was required created other risks, including road traffic fatalities, air pollution and climate change. Another example is genetically modified organisms (GMOs), which could possibly eradicate world hunger, yet may potentially threaten biological diversity and sustainability as well as traditional agriculture as practiced by socio-economically challenged farmers globally (e.g. Robaey, 2015).

In addition to, or perhaps because of its complex relation to risks, technological design has a complex link to concerns about responsibility. Technological design and innovation potentially affect responsibility in different ways. First, technologies can affect the distribution and balance of responsibility between individuals. Second, as will become clear in Chapter 6, technologies can alter the way in which responsibility is distributed between individuals in different roles, such as parent and child. Third, as in the example of alcohol interlocks described in Chapter 2, new technology can affect the way in which responsibility is distributed between individual and collective agents, such as the government and the industry. Finally, technology potentially affects our willingness to *take* responsibility, i.e. our behaviour over time. In this chapter a different question concerning responsibility and technological design is explored: Who is responsible for technological design and its consequences? Responsibility as a virtue is analysed in relation to this question concerning responsibility for technological risk-creation.

Risky technologies

I would, and have, argued that taking part in risk-creating activities entails moral responsibility (e.g. Van de Poel and Nihlén Fahlquist, 2012). Technological innovation, even though generally highly beneficial to humankind, is a risk-creating activity. Engineers commonly take part in such activities. They create tremendous opportunities for people and the environment, they contribute to societal progress and they make life easier and safer for people. However, their actions and decisions and the projects in which they play a role also create risks to people's lives as well as to the environment and to future generations. What should we, as a society, do about those risks? Should society adopt the precautionary principle and consider slowing down innovation and technological and scientific progress? That does not appear to be a viable or an ethically legitimate option. As Sandin and others illustrate, to not do anything, to maintain the status quo and to remain passive is also a risky choice with potentially ethically troublesome consequences (Sandin, 2004).

Van de Poel (2016) and others have used the concept of social experiment to describe the situation when new technology is introduced to society with considerable uncertainty, e.g. nuclear energy or GMOs. It is similar to a medical experiment in some sense, he argues, but without the controlled environment of the medical laboratory. The argument is based on the idea that we cannot do without these social experiments, but that they should still be done responsibly. The notion of a social experiment would rephrase the question of whether a technology is acceptable or not as a matter of how, and under what conditions, the experimentation would be morally responsible (Van de Poel, 2011).

Peterson questions the introduction of the concept of social experiment, arguing that the concept does not help answer the core question: Is technology X ethically acceptable? (Peterson, 2013). Regardless of whether we call it social experiment or merely discuss the technology as such, emerging technologies bring threats as well as opportunities for humanity, and the introduction of new technologies bears some similarity to medical experiments. Stating that technological development should be carried out responsibly is hardly controversial. However, when it comes to specifying what this means or who, exactly, is responsible for technological developments and the risks it creates, things become more controversial and complicated. I argue that not only are managers, politicians and leaders responsible for technological development and the risks it entails, but each and every engineer taking part in such development is also responsible. This does not mean that they are all equally to blame when things go wrong. However, they do have a forward-looking moral responsibility to do what they can, within reasonable limits, to contribute to the responsible design and use of the results of their work. In my view, the best way to conceive of such responsibility is the virtue-ethical notion introduced in this part of the book. Unless engineers develop a "sense" – a virtue – of responsibility, it is difficult to see how technological development, which is extremely powerful in modern society, can be a beneficial activity for our society. The need to introduce the notion of responsibility as a virtue is due in part to the complex and collective setting of modern engineering. I will analyse the notion of responsibility as a virtue in relation to the so-called problem of many hands, one of the greatest risks to responsible engineering today. However, let us start with a current example of how technological development carries risks as well as benefits.

Example: autonomous cars and ethical responsibility

Autonomous, or driverless, cars are the result of technology that promises to radically reduce the number of fatalities and injuries due to road traffic. This would of course be tremendously beneficial to people and society. Nevertheless, there are risks involved in delegating moral and legal responsibility, in terms of control for traffic safety to technology. In March 2018, a woman was killed in Arizona by an autonomous car (Wakabayashi, 2018). The cause of the accident is not exactly clear, but what we know at this point by our discussion is that causation is not as simple as one might think. Autonomous cars make the so-called philosophical thought experiments called "trolley cases" highly relevant. There are many different versions of this thought experiment, but in the original trolley case, we are told to think about a situation in which we see a group of five people standing on the railway tracks. There is a runaway trolley going in the direction of the group. If we pull a lever in front of us, we will be able to direct the trolley onto a different track where only one individual is standing. Because this is a philosophical and hypothetical example created to confound our moral intuitions, all other things are the same in both cases. In other words, in this case, there is nothing we can do, such as yelling to the people to get them to realise the threat, and we know for sure that either the whole group of five or the one person will be killed. The question is then whether we would, and possibly should, pull the lever and save five people while letting the lone individual die (e.g. Foot, 1978; Kamm, 1989). Most people answer affirmatively, thereby pointing towards a utilitarian conclusion that it is morally permissible and possibly even obligatory to sacrifice one in place of a group of people. The puzzle for moral philosophers presents itself when this experiment is compared to a similar one, in which we are supposedly standing on a footbridge above the tracks, meaning that instead of pulling a lever, we could push a large man onto the track to save the five, although this will kill the large man. Most people would hesitate to do that and would perhaps even say that it would be highly immoral to do so. How are the opposing intuitions in these two experiments to be reconciled? This puzzle has been discussed by philosophers and investigated by social scientists (e.g. Greene *et al.*, 2001).

Autonomous cars shed light on the importance of trolley cases. Of course, if these cars would eliminate rather than cause fatalities, the thought experiment is not specifically relevant to this technology. However, if autonomous cars are introduced on a large scale into our traditional infrastructure, chances are high that they will, at some point, must, "choose" one road user over another, a group over an individual, or an elderly person over a child.

If one must choose between five anonymous people and one anonymous person, the answer is fairly uncomplicated; but it is easy to change the experiment slightly in such a way as to increase our hesitation to pull the lever. What should the car "decide" to do if it has to "choose" between one four-year old child and a group of 70-year olds? What should it do if it is presented with the "choice" between killing two slightly intoxicated pedestrians crossing a street or the driver herself?

In my view, the case of autonomous cars raises three groups of ethical questions:

1 How should the cars "decide" in situations where one or more people might die or be injured as a consequence?
2 Who is morally (and legally) responsible for the "decisions" and "actions" made by autonomous cars?
3 What are the long-term societal consequences of introducing autonomous cars on a large scale?

Is there anyone who would claim that these questions are not important or even irrelevant? If they are relevant, are perhaps engineers not the ones who should address these questions? It could be argued that engineers should just be focused on engineering, possibly as a creative and problem-solving activity, and that the ethical and societal discussion should come in at a later point. It could be argued that ethical analysis would slow down this creative and problem-solving process, assuming that slowing down is a negative thing. In 2011, I contacted a CEO for a company developing GPS technology that would be used to track children, and I asked him whether they had considered any of the ethical issues related to their technology. Even though ethics of technology is a research area of its own, his answer is typical of the way society in general views technological innovation. "We are not there yet", he said. "At this moment, we are merely focused on getting the product on the market. We would be willing to take up the ethical discussion later." I discuss the ethical aspects of tracking children in Chapter 6, so I will not do that here. However, it is interesting to note that companies tend to think that ethics is something to be discussed after the fact, that is, after the product exists. In contrast, ethics of technology research illustrates the need to reflect ethically about technology during the development phase, for example through value-sensitive design (e.g. Asveld and Roeser, 2009; Nissenbaum, 2016; Van den Hoven *et al.*, 2015).

Furthermore, the societal image of the engineer is of an objective and neutral figure and including ethical values may be seen as a challenge to this image. Engineers, Roeser argues, are seen as the "archetype of rational and quantitatively oriented people" (Roeser, 2012). However, it is difficult to find researchers who would argue this way officially, that is, who would explicitly state that engineers do not need to engage in ethical issues. This might mean that there are very few researchers today who would criticise the notion that engineers have an ethical responsibility. Alternatively, it could mean that few people would make this argument publicly. Nevertheless, I think there is adequate academic consensus around the suggestion that engineers have a responsibility to include ethical values and concerns in their projects, and there are a number of books attempting to teach engineering students how to reflect on ethical issues (e.g. Davis, 1998, 2005; Hansson, 2017; Van de Poel and Royakkers, 2011).

However, what exactly is implied by the normative statement that engineers should take ethical values into account, and how far does that responsibility reach? When trying to delineate such a responsibility in practice, there may be more disagreement than there is in theory. It could be argued that autonomous cars save more lives and so, from a utilitarian perspective, are preferable to traditional cars. This may be seen to imply that we do not need to further discuss any other ethical concerns. One question is whether the vehicle industry, against this background, needs to be involved in the discussion about the ethics of autonomous cars. Do they really have to engage in ethical issues when the whole point of the technology is to save lives and to reduce the general level of risk associated with road traffic? First, it should be acknowledged that there is currently uncertainty regarding the changes that large-scale autonomous cars would entail. Furthermore, there may be cases where autonomous cars save a group of people while harming one individual. Even though this would be right according to many, and certainly according to a strict interpretation of utilitarianism, it might still be ethically problematic. Even though it is less bad to cause one death than to cause many, the death of even one person is still potentially problematic. A human being who is forced to avoid a greater harm by causing a lesser harm, such as killing one individual to save a group of people, is likely to be emotionally affected even if she could not have acted in a less blameworthy way in that specific situation.

Responsibility as a virtue presupposes an emotional capability to care about the potential victims of one's activity. As described by philosophers, emotions are necessary for moral reflection (e.g. Nussbaum, 2001). This is true for moral reflection about risky activities as well, as pointed out by Roeser (2010). As Little argues, the emotional aspect of morality goes even deeper. Little argues that "seeing is caring", in other words, that truth is sometimes clearer from a position of "sympathetic thinking" or "appropriate affect" (Little, 1995). Roeser and Little challenge the traditional idea that morality requires adopting a distanced role and using the detached faculty of Reason presupposed in traditional Western philosophy. The notion that emotions are not completely distinct from Reason has become influential during the last decades, in philosophical debate as well as in neuroscientific and psychological research (Blum, 1994; Damasio, 1994; de Sousa, 1987; Goldie, 2000; Greenspan, 1988; Little, 1995; Nussbaum, 2001; Roberts, 2003; Solomon, 1993; Stocker, 1996).

Emotional capability is important in the context of autonomous vehicles and trolley cases. In some cases, it may be a good thing to feel regret that one has caused a death or injury even though one did everything one could. The role of moral regret has been explored by Williams. He argues that even in cases where we cannot honour two conflicting "oughts", the fact that we choose one does not resolve the conflict, but rather leaves a "moral remainder" or "moral regret" (Williams, 1973).

The ethical analysis of autonomous cars could benefit from similar discussions concerning the use of robots in modern warfare. Autonomous cars are similar to such automated robots in some ways, in that (a) they can both be used in ways that end up killing or injuring someone, (b) they can both be used on a large scale in which they could save more lives than if they were not used and (c) they can be used without human control in action but with human control in design. They are different with regard to the context in which they are employed and the intention with which they are developed. Whereas autonomous cars are developed in a civil context in which the overall purpose is to transport people without harming anyone, military robots like drones were developed in a military context where the primary aim is not to do no harm, but perhaps to do proportional or limited harm, given the purposes of the conflict. As has been argued in the context of automated robots used in wars, a decision to take a human life "must be an inherently human decision" and that it is "unethical to allow a machine to make such a critical choice" (Johnson and Axinn, 2013). I do not wish to take a firm stand on whether I agree and whether decisions that can lead to fatality or injury always have to be made by humans, or whether they should always be made through the lens of emotions. However, I think that we do not currently have enough knowledge about either the role of emotions in decision-making or the consequences of letting non-human agents make such decisions. This lack of knowledge and certainty should not cause us to put an end to technological development in this field, but it should promote reflection and perhaps slow development somewhat. Perhaps it would be a beneficial technology with a certain measure of reflection and perhaps even humility.

Now that we have an idea about how and why responsibility matters in the context of technological development and risk, I will look more closely at a challenge to responsibility in engineering, i.e. the so-called problem of many hands (PMH). It will be argued that responsibility as a virtue is essential to counteract this problem, which is caused by the complexity of technological risk-creation in modern society.

The problem of many hands[1]

Arguably, collective responsibility is more important now than it has ever been. Human activities increasingly take place in collective contexts. Research, innovation and technological development are often the result of large projects that involve many different actors. When a large number of agents are involved in causing a negative outcome, it may be hard to identify who is responsible. This has been referred to as the "problem of many hands" (Thompson, 1980; Van de Poel *et al.*, 2012). It was first formulated in a political context, referring to the moral responsibility of public officials. Thompson argues that because so many different officials contribute to policies in many different ways, it is difficult to ascribe moral responsibility to anyone specific for the organisation's conduct. For the public, it is almost impossible to find a single individual who can be said to have independently made some crucial decision that led to the undesired situation, that is, an individual who caused it knowingly and freely. The problem of many hands is now primarily discussed in engineering ethics and business ethics (Van de Poel *et al.*, 2012).

The problem of many hands can be seen as an epistemic problem, i.e. a problem of how to identify the person who was causally responsible, who "did it". Insiders usually know more about causal links, so this could be seen mainly as a problem faced by outsiders (Davis, 2012). Construed in this way, the problem could be avoided by making each individual's causal contribution more transparent. To introduce more transparency might not be such a bad idea, but it would not solve the entire problem. Rather, the problem is an *ethical* problem if interpreted as a situation in which none of the individuals is (or can be held) *morally* responsible, but in which the collective of individuals is *morally* responsible (Van de Poel *et al.*, 2012). Consider, for example, a situation where autonomous cars are introduced on a large-scale basis in a country without having been properly tested. This is presumably what van de Poel would call a social experiment (Van de Poel, 2011, 2016). Imagine that the experiment involves introducing autonomous vehicles without having tested them to the same extent that a new drug would have been tested. Let us assume that these cars, which are de facto decision-makers in road traffic, start injuring small children and dogs in order to avoid damage to the car and injuries to the driver. After all, vehicle-producing companies currently have to take marketing potential into consideration when designing new cars. One question is whether it would possible to market a car that might "sacrifice" the driver in certain situations. Who is responsible for the children and dogs whose lives may be lost? There are multinational companies involved, the industry as such, managers, programmers, project managers, investors and politicians as well as public officials at the national and international transportation agencies. All of these together potentially brought this situation about, this situation in which lives are lost. The question of responsibility is partly about backward-looking responsibility in such a scenario: Who caused this unfortunate, to say the least, situation? However, it is also partly a question about forward-looking responsibility in the sense discussed in this chapter: Who should take responsibility now in order to prevent such a scenario from happening in reality? Arguably, engineers taking part in the development and programming of these vehicles need to take responsibility in order to prevent such situations from occurring.

The moral responsibility of engineers can be seen merely as a matter of forward-looking task distribution, that is, as a dictation of the results that should be achieved, such as "a safe and environmentally friendly car". Additionally, as I argue in this chapter, it could be conceived as a virtue, that is, a character trait or a disposition, to take responsibility for more than what one's employment contract specifies. Arguably, in order to avoid ending up in a situation where no one is responsible for an unwanted situation in which lives are lost and the environment is severely threatened, we need people involved in risk-creating activities to actively take responsibility, and we need organisations to promote responsibility conceived as a virtue and to encourage engineers to cultivate such a virtue.

"Responsible" organisations[2]

How can a situation be achieved in which people are more responsible in the virtue-ethical sense described above? If responsibility as a virtue consists of care, moral imagination and practical wisdom, what does that mean in the context of collective actors, e.g. organisations and networks? The challenge when applying the concept to such institutional contexts is that it does not immediately make sense to apply virtues to a group of people. It appears that care is primarily a personal quality. Individuals are more or less caring and empathetic, but it seems awkward to talk about caring organisations or caring networks (other than in commercials). Care and empathy should probably be seen as individual qualities, that is, as a combination of certain character traits, attitudes and emotions. However, the other two components of responsibility as a virtue – moral imagination and practical wisdom – could more easily be applied to the collective context. Moral imagination can probably be practiced and improved if time and energy is devoted to collectively detecting risks and to discerning the ethically relevant aspects of an activity. Practical wisdom is the ability to deliberate about values and to transform the concerns raised through care and moral imagination into actions and behaviour over time. This is also something that possibly can be facilitated and made more systematic by institutional means. It would probably require paying attention to and rewarding individual members of the larger organisation for their caring concerns, attitudes, emotions and combined moral imagination, as well as finding ways to make sure these attitudes have appropriate effects on the activities performed in the organisation or network.

Upbringing, education and peer influence as well as genes contribute to creating people's personalities. For such reasons, we may not be able to change people's personalities. What, nevertheless, can be done to create a situation in which behaving responsibly is possible and encouraged, i.e. an institutional environment conducive to responsible behaviour? Obstacles can be removed, systems of incentives and rewards for good behaviour can be created and organisations can make it easier to do good in the context of large development projects, all of which would counteract the image of the engineer or manager as a cog in a cogwheel.

I will give two illustrations of how the organisational context could be shaped to encourage or discourage responsibility as virtue. The first example shows how choices can be affected. As we have seen in previous chapters, Thaler and Sunstein (2008) have described how all choices are affected by the way in which alternatives are presented and argue that even a libertarian needs to be paternalistic in this sense: there are no choices completely unaffected by the number and framing of alternatives. The person who contextualises a choice in this way is called a "choice architect". One of their basic examples is a cafeteria where the manager needs to decide how to present the food products sold: fruit, candy, food and so forth. One decision would be to make the candy more prominent in order to encourage people to buy a lot of candy. Another option, arguably a more responsible option, is to put the fruit in the more visible place and the candy in a less visible place. This way, the manager has not removed any of the options, but she still potentially contributes to people making healthier choices.

This example illustrates how a social or institutional context conducive to (ir)responsible behaviour can be created. Virtue ethicists argue that habituation and moral exemplars can make people more virtuous. If it becomes the norm for engineers, managers and the like to discuss and always think about potential risks and ethical aspects of the activities they are involved in, and if they get used to this norm, it will become a habit. If leaders talk and act in a responsible manner, i.e. if they care, use their moral imagination and transform their emotions and attitudes into action, this is likely to have a positive influence. And if people are habituated and "programme" themselves into acting responsibly, then they are more likely to do it even when there is a cost. In every network there are leaders or initiators, and if these people make sure good and responsible behaviour is rewarded instead of punished, this too is likely to have some effect. If "choice architects" make it easier to choose the responsible behaviour than the irresponsible behaviour, that will influence people's behaviour.

To stimulate responsibility as virtue in collective settings, the first thing to do would be to incorporate ethics into the practices of organisations in a more profound way than what is common today, i.e. not merely as a matter of PR. In agreement with my suggestion in this chapter, it has been acknowledged in the context of business ethics that organisations should focus less on codes of conduct and regulation and instead emphasise the role of virtues, character and attitudes (Clegg and Rhodes, 2006). An organisation does not become ethical by establishing rules that aim at preventing wrongdoing by its employees. For an organisation to become an ethical organisation, it must consist of people with a high moral standard, who aspire for excellence, who do good things and who care both about doing a great job as well as about other people. This obviously means that the most excellent people are to be recruited. However, in the context of the power and risk-generating nature of technological innovation, excellent *and* responsible people should be recruited. Responsible engineers do not see themselves merely in their professional role and do not compartmentalise but attempt to see the whole instead of merely the parts or details. People who lack caring attitudes and emotions should perhaps not even work in risk-generating activities. From an ethical point of view it may be as simple as that. Unless one cares and has the emotional ability to be concerned with how the activities in which one participates create risks to people's health, lives and the environment, one should not be entrusted with such power.

In order to achieve a situation in which engineers are ethically aware and emotionally/morally sensitive, we need not only to focus on recruiting the right people, but also on training them to develop these skills and character traits. Character traits are of course partly a matter of genes and upbringing, but also of habituation. Managers and leaders are obviously important as well and can serve as moral exemplars.

In addition to recruiting virtuous/responsible/caring people, other things can be done to develop ethical and emotional sensitivity in employees, such as providing ethics courses and discussions of scenarios and cases. The general aim of both recruiting the right people and continuing to develop their attitudes is to establish an organisational culture conducive to ethical behaviour, which includes doing good and having a caring attitude in addition to the avoidance of wrongdoing. Kaptein (2011) has shown that the ethical culture of an organisation strongly influences the ethical behaviour of employees. For example, management can reduce unethical behaviour by creating an organisation in which managers and supervisors act as ethical role models. Unless managers themselves are responsible and caring, it is difficult to see how they would be able and willing to create an organisational culture characterised by virtuous/responsible behaviour and a caring attitude.

Avoiding the problem of many hands – towards responsible technological development

We have seen that responsibility as a virtue could be conceived of as involving three main character traits: care, moral imagination and practical wisdom, and we have seen how institutional contexts may be shaped to be conducive to responsibility as virtue. Let us now turn to the question of how responsibility as virtue can contribute to solving or avoiding the PMH.

Although all normative meanings of responsibility are important for the PMH, there is one way in which responsibility as a virtue is different from especially the backward-looking connotations of responsibility. In cases of backward-looking responsibility, an agent is usually *held* responsible, and for this attribution of responsibility to be fair, certain conditions must apply. If these conditions are not met, it will – at least in normal circumstances – not be considered fair to hold an agent responsible in a backward-looking sense. In contrast, responsibility as virtue does not refer to holding responsible but to *taking* responsibility. Although there are limits to which responsibilities agents can (rationally) take, agents can take responsibility for issues for which they cannot be reasonably held responsible by others.

When we look at forward-looking responsibility, it could also be argued that responsibility conceived as moral obligation is often taken or voluntarily assumed by an agent rather than attributed by others. Still, there are important differences between both meanings of forward-looking responsibility. One of the things that distinguishes virtue from obligation is that virtuous agents will assume responsibilities conceived as moral obligations even when this is not strictly morally required. In other words, virtuous agents now and then act in a *morally supererogatory* way, i.e. beyond what is strictly morally required or can be demanded by others. In the deontological approach, agents can also assume moral obligations beyond what is required from them by making a promise or commitment; however, in that case the additional responsibilities need not be morally desirable, they should just be morally acceptable (i.e. they can be morally neutral). Typically, responsibility as virtue encourages an agent to take on moral obligations that are morally desirable but not strictly required.

There is also another factor that sets responsibility as a virtue apart from moral obligations. Obligations are usually relatively specific: they refer to specific states-of-affairs to be achieved or avoided. To take on such moral obligations, one first needs to be aware of what is morally at stake in a situation. This often requires moral imagination and care, which, as we have seen, are two of the main elements of responsibility as a virtue. Responsibility as a virtue thus is conducive to an awareness of moral issues, which is a precondition for identifying the relevant moral responsibilities as moral obligations that are to be assumed (or attributed).

How does encouraging responsibility as a virtue avoid the problem of many hands? There are two ways in which responsibility as a virtue might do so. First, as argued above, virtuous agents will assume more responsibilities as moral obligations or tasks and will better see what kinds of obligations or tasks are desirable. So, in an organisation or network that is conducive to responsibility as a virtue, more agents will assume more responsibilities as moral obligations. This will ideally reduce the occurrence of the PMH with respect to responsibility as obligation.

Additionally, responsible engineers take ownership of the collective results of their company's activities, i.e. for more than their isolated work product. As Michael Davis (2012, p. 22) notes: "I can take responsibility for what happened even if I have no idea how I contributed to the ultimate decision (as, for example, President Truman did when he said, 'The buck stops here')". As Davis suggests, we can take responsibility for things for which others cannot hold us responsible. This can include the responsibility to do better in the future, which is primarily a forward-looking responsibility, but it will usually also include an element of accountability or blameworthiness, because it requires an analysis of what went wrong in the past to know how to avoid harm in the future. Analysing what when wrong in the past seems to imply, at least de facto, assuming responsibility as accountability.

Risk detection and management

I have argued that engineers who partake in risk-creating activities should develop responsibility as a virtue, and that organisations should promote this kind of character trait. In this context, the question as to how responsibility as virtue can help address technological risks with regard to detection and prevention, and with how to deal with them when they materialise.

As mentioned previously, Roeser (2012) argues that it is highly important that engineers who partake in risky activities use their intuition and emotional abilities to detect risks. Engineers should use their concerns in the design of their research and technologies, for example by building barriers to prevent certain hazards from occurring or by in some cases applying the precautionary principle, meaning that technologies whose consequences are hard to predict should first be investigated in a safe setting. The precautionary principle should also be applied to new technologies with potential hazards when there is scientific controversy or uncertainty about these hazards. One technology to which the precautionary principle has been applied is nanotechnology, because experts have great worries about its potential risks. These worries can "point to a source of danger to our well-being" (Roeser, 2012, p. 109).

In order to achieve this goal, ethical reflection should be taught and practised at universities. Engineering students should therefore learn to use their imaginative and emotional skills at engineering universities. This could be compared to medical doctors' degree programmes, where students practice ethical analysis throughout their education so that they will make better decisions once they become practicing physicians. This means that ethics modules for engineers should not merely be focused on reasoning skills and argumentation, but also on developing sympathy and similar emotions in order to raise awareness about the importance of values in design. This does not mean that decisions should only be based on emotions or that engineers who use emotions will necessarily make ethical or responsible decisions. As Le Coz and Tassy explain in relation to medical decision-making, emotions are necessary but not sufficient for ethical decision-making (Le Coz and Tassy, 2007).

I agree with this view and would add that the emotions involved in care are also necessary for engaging in responsibility as a virtue. This implies that unless someone has this emotional capacity they are unable to exercise this kind of responsibility, and that unless someone is responsible, that person should not be allowed to engage in activities that generate risks to health and the environment. Whereas people's upbringing cannot be influenced, what can be done is to include these elements when educating people who will be involved in this kind of activities, as suggested by Roeser (2012).

In ethical parallel research about the development of a new sewage treatment technology undertaken by Zwart and van de Poel (2006), it was observed that the researchers and users did not consider themselves responsible for the potential so-called secondary emissions of sewage treatment plants, such as hormone-related substances, heavy metals and pathogenic bacteria. They believed that secondary emissions were likely to not be a major risk factor, and whereas the researchers allocated the responsibility for investigating and preventing secondary emissions to the users, the latter generally thought it was the researchers' responsibility to investigate and prevent such emissions. After this observation by the ethicists was reported back to the technological researchers, some began to take the potential risk of secondary emissions more seriously and they consequently applied for funding to explore the risks of secondary emissions (De Kreuk *et al.*, 2010). This shows that there is a lot of conventional and habitual thinking involved in innovation, and that once the actors are encouraged to reflect and to discuss risks and responsibility for risks, they may assume responsibility for risks that they initially did not consider their responsibility. Habitual thinking can thus be changed. If efforts are made to make members of innovation networks more prone to detect risks, their ability to manage risks is expanded.

A second example from a trailer design project shows that engineers developing the trailers plainly did not see it as their responsibility to do anything beyond what the customer had decided they wanted. Traffic safety, for example, was not considered part of their responsibility. Even if there were measures they could take to improve safety, they did not do so; they did not even suggest it to the customer (van de Burg and van Gorp, 2005). The point of bringing up this example is not to blame the engineers. It shows how we habitually see engineers as solving technical problems in an amoral way. This could and should be changed. A responsible engineer takes responsibility for important values, for doing their utmost to ensure the safety of those people who will be inside or outside of the vehicle they participate in designing. If engineers get used to using their emotional capacity and to developing their moral sensitivity, they will ideally care about the potential victims of the risks their activities generate and use their imaginations and their practical wisdom to avoid unnecessary risks.

The next step would require that everyone take responsibility for managing risks. This does not mean that everyone is equally responsible, but that everyone is sufficiently concerned to make sure tasks are distributed. When everyone involved cares, is able to imagine together what potential risks there are and to take responsibility for creating an accurate and inclusive agenda of risks, responsibilities in terms of tasks can be distributed. In Chapter 1, we discussed the importance of distributing responsibilities in a fair way.

In situations where risks are assessed through a cost-benefit analysis, a decision is made to accept the risk when the benefits outweigh the costs or risks and vice versa. Psychologists have shown that lay people do not merely calculate risks based on costs and benefits but include more qualitative concerns such as whether the risk is taken voluntarily, whether the benefits and costs are distributed fairly, whether the consequences might be disastrous and so forth. Ethicists have shown that many, if not all, of these concerns are also normatively justified (cf. Roeser). Therefore, these other concerns should be taken into account by people who work in risky innovation projects.

Is this demanding too much?

Some may object to the suggestions in this chapter. They may think this is asking too much from individual engineers. One way of thinking about this question of whether it is too demanding to expect engineers to care in this emotional sense of the word, is to adopt Slote's view on what he calls "balanced care". He argues that we should strive for balance among three kinds of caring: self-concern, intimate caring, and humane or humanitarian caring:

> Concern with intimates as a class and excluding the agent must be balanced against concern for (the rest of) humanity. This much is required. But then it is merely permissible that self-concern be in balance with each of these other concerns. That is, someone whose actions express or reflect some sort of balance among self-concern, intimate caring, and humane caring acts rightly or permissibly, but s/he may also act permissibly and will indeed act supererogatory if her actions reflect a *lesser* degree of self-concern than just mentioned.
>
> (Slote, 2001, p. 78)

If this qualification is applied to collective settings, it means that an organisation is allowed to have several affiliations and obligations. What is characteristic of a responsible engineer, that is, of someone who cares and has moral imagination and practical wisdom, is that these cognitive-affective abilities facilitate their capability for responding to a plurality of normative demands and for knowing how to prioritise them.

Conclusion

Responsibility as virtue can be seen as consisting of care, moral imagination and practical wisdom, and this involves having a certain attitudinal-emotional concern and the ability to transform those concerns into actions and responsible good behaviour over time. Part of the explanation for why some people are responsible in this sense whereas others are not is due to genes and upbringing, that is, to factors we can do little about when dealing with these issues on a societal level. Nevertheless, as we have seen, there are various things we can do, or at least try to do, in order to stimulate responsibility as a virtue in collective contexts. We can shape institutional settings to make them more conducive to responsibility-taking. We can hire people in organisations who are not just excellent professionals but who also exhibit responsibility as virtue, and finally, we can train people to acquire the competences and attitudes that are required for implementing responsibility as a virtue.

In this chapter, I have argued that responsibility as a virtue helps to overcome or at least diminish the problem of many hands in the context of engineering and technological risk. As we have seen, there are several reasons why responsibility as a virtue might help to reduce the problem of many hands. The main underlying reason, however, seems to be that responsibility as a virtue is not so much about holding others responsible or attributing responsibility to others, but about actively taking responsibility. Even in situations in which it may be inappropriate or too strict to hold another responsible, one might choose to take responsibility oneself.

Responsibility as virtue requires a certain attitude. Part of this attitude requires that people who work in collective settings in which potential harm is created see themselves not merely as a cog in a cogwheel, but as a part of a larger societal context. Everyone should ideally feel responsible for the consequences of the activities of the collective. This does not mean that each and every person is equally blameworthy when risk materialises. I do not wish to suggest that everyone should be blamed when things go wrong as a consequence of a collective project. However, it may be a positive thing that people sometimes feel ashamed, experience regret or become morally tainted by what happens as a consequence of collective activities even if they did nothing to directly cause it. This may be proof of the fact that someone who cares about actual and potential victims can imagine what it feels like to be in their shoes, which will lead her to do what she can to reduce the negative effects of what happened. In short, an engineer's emotions of shame or regret can be seen as evidence that she feels responsible for what the collective of which she is a part does. And that would be in order.

Notes

1 This section is based on Nihlén Fahlquist, J. 2017. Responsibility analysis. In S.O. Hansson (ed.), *Ethics of Technology: Methods and Approaches* (Chapter 9). London: Rowman and Littlefield.
2 These sections are based on Nihlén Fahlquist, J. 2015. Responsibility as a virtue and the problem of many hands. In I. van de Poel, L. Royakkers and S. Zwart (eds), *Moral Responsibility and the Problem of Many Hands* (Chapter 6). London: London.

References

Asveld, L. and Roeser, S. (eds). 2009. *The Ethics of Technological Risk*. London: Earthscan.
Blum, L.A. 1994. *Moral Perception and Particularity*. New York: Cambridge University Press.
Clegg, S.R. and Rhodes, C. 2006. Introduction: questioning the ethics of management practice. In: S.R. Clegg and C. Rhodes (eds), *Management Ethics: Contemporary Contexts* (pp. 1–9). London: Routledge.
Damasio, Antonio R. 1994. *Descartes' Error: Emotion, Reason, and the Human Brain*. New York: Putnam.
Davis, M. 1998. *Thinking Like an Engineer: Studies in the Ethics of a Profession*. New York: Oxford University Press.
Davis, M. 2005. *Engineering Ethics*. London: Routledge.
Davis, M. 2012. Ain't no one here but us social forces: constructing the professional responsibility of engineers. *Science and Engineering Ethics* 18(1): 13–34.
De Kreuk, M., Van de Poel, I., Zwart, S.D. and Van Loosdrecht, M.C.M. 2010. Ethics in innovation: cooperation and tension. In: I. Van de Poel and D.E. Goldberg (eds), *Philosophy and Engineering: An Emerging Agenda* (pp. 215–226). Dordrecht: Springer.
De Sousa, R. 1987. *The Rationality of Emotions*. Cambridge, MA: MIT Press.
Foot, P. 1978. The problem of abortion and the doctrine of the double effect. In *Virtues and Vices and Other Essays in Moral Philosophy*. Oxford: Basil Blackwell (originally appeared in the *Oxford Review*, Number 5, 1967).
Goldie, P. 2000. *The Emotions: A Philosophical Exploration*. Oxford: Oxford University Press.
Greene, J.D., Sommerville, B., Nystrom, L.E., Darley, J.M. and Cohen, J.D. 2001. An fMRI investigation of emotional engagement in moral judgment. *Science* 293(5537): 2105–2108. DOI: 10.1126/science.1062872.
Greenspan, P. 1988. *Emotions and Reasons: An Inquiry into Emotional Justification*. New York/London: Routledge.
Hansson, S.O. (ed.). 2017. *The Ethics of Technology: Methods and Approaches*. London: Rowman & Littlefield.

Johnson, A.M. and Axinn, S. 2013. The morality of autonomous robots. *Journal of Military Ethics* 12(2): 129–141.

Kamm, F.M. 1989. Harming some to save others. *Philosophical Studies* 57(3): 227–260.

Kaptein, M. 2011. Understanding unethical behavior by unraveling ethical culture. *Human Relations* 64(6): 843–869.

Le Coz, P. and Tassy, S. 2007. The philosophical moment of the medical decision: revisiting emotions felt, to improve ethics of future decisions. *Journal of Medical Ethics* 33(8): 470–472. DOI: 10.1136/jme.2006.016741.

Little, M.O. 1995. Seeing and caring: the role of affect in feminist moral epistemology. *Hypatia* 10: 117–137.

Nihlén Fahlquist, J. 2015. Responsibility as a virtue and the problem of many hands. In I. van de Poel, L. Royakkers and S. Zwart (eds), *Moral Responsibility and the Problem of Many Hands*. London: Routledge.

Nihlén Fahlquist, J. 2017. Responsibility analysis. In S.O. Hansson (ed.), *Ethics of Technology: Methods and Approaches*. London: Rowman and Littlefield.

Nissenbaum, H. 2016. *Values at Play in Digital Games*. Cambridge, MA: MIT Press.

Nussbaum, M.C. 2001. *Upheavals of Thought: The Intelligence of Emotions*. Cambridge: Cambridge University Press.

Peterson, M.B. 2013. New technologies should not be treated as social experiments, *Ethics, Policy & Environment* 16(3): 349–351. DOI: 10.1080/21550085.2013.844574.

Robaey, Z.J. 2015. Looking for moral responsibility in ownership: a way to deal with hazards of GMOs. *Agricultural and Environmental Ethics* 28(1): 43–56. https://doi.org/10.1007/s10806-014-9517-8.

Roberts, R.C. 2003. *Emotions: An Essay in Aid of Moral Psychology*. Cambridge: Cambridge University Press.

Roeser, S. 2010. *Emotions and Risky Technologies* [Electronic resource]. Dordrecht: Springer Netherlands.

Roeser, S. 2012. Emotional engineers: toward morally responsible design. *Science and Engineering Ethics* 18(1): 103–115.

Sandin, P. 2004. *Better Safe than Sorry: Applying Philosophical Methods to the Debate on Risk and the Precautionary Principle*. Diss. Stockholm: Tekn. högsk.

Slote, M. 2001. *Morals from Motives*. Oxford: Oxford University Press.

Solomon, R. 1993. *The Passions: Emotions and the Meaning of Life*. Indianapolis, IN: Hackett.

Stocker, M. (with E. Hegemann). 1996. *Valuing Emotions*. Cambridge: Cambridge University Press.

Thaler, R. and Sunstein, C. 2008. *Nudge: Improving Decisions about Health, Wealth, and Happiness*. New Haven, CT: Yale University Press.

Thompson, D.F. 1980. Moral responsibility of public officials: the problem of many hands. *American Political Science Review* 74: 905–915.

Van de Poel, I. 2011. Nuclear energy as a social experiment. *Ethics, Policy & Environment* 14(3): 285–290.

Van de Poel, I. 2016. An ethical framework for evaluating experimental technology. *Science and Engineering Ethics* 22: 667. https://doi.org/10.1007/s11948-015-9724-3.

Van de Poel, I. and Nihlén Fahlquist J. 2012. Risk and responsibility. In S. Roeser, R. Hillerbrand, M. Peterson and P. Sandin (eds), *Handbook of Risk Theory*. Dordrecht: Springer.

Van de Poel, I. and Royakkers, L. 2011. *Ethics, Technology, and Engineering: An Introduction*. Chichester: Wiley-Blackwell.

Van de Poel, I., Nihlén Fahlquist, J., de Lima, T., Doorn, N., Royakkers, L. and Zwart, S. 2012. The problem of many hands: climate change as an example. *Science and Engineering Ethics* 18: 49–67.

Van den Hoven, J., Vermaas, P.E. and van de Poel, I. (eds) 2015. *Handbook of Ethics, Values, and Technological Design: Sources, Theory, Values and Application Domains*. Dordrecht: Springer Netherlands.

van der Burg, S. and van Gorp, A. 2005. Understanding moral responsibility in the design of trailers. *Science and Engineering Ethics* 11(2): 235–256.

Wakabayashi, D. 2018. Self-driving Uber car kills pedestrian in Arizona, where robots roam. *New York Times* 19 March.

Williams, B. 1973. Ethical consistency. In B. Williams, *Problems of the Self: Philosophical Papers 1956–1972*. London: Cambridge University Press.

Zwart, S.D., Poel, van de, I.R., Mil, van, H.G.J. and Brumsen, M. 2006. A network approach for distinguishing ethical issues in research and development. *Science and Engineering Ethics* 12(4): 663–684.

5 Responsible risk communication

Introduction

Risk communication holds a central position in contemporary public debate. Risks are communicated by government agencies, mass media and social media. We are told to avoid certain behaviour, to adopt certain habits and to do certain things. The intentions are generally good, aimed at prolonging our lives and increasing our safety and well-being. However, even when intentions are good – or perhaps even more so in such cases – we should use our critical lens to analyse the ethical aspects and consequences of policy.

As mentioned in previous chapters, research findings in risk perception have shown that experts and lay people perceive risks very differently. Whereas the former make decisions concerning the acceptability of risks based on calculations regarding probability and outcome, lay people have a more complex and nuanced view. When judging whether a certain risk is acceptable, most people take into consideration, for example, whether the risk is voluntarily taken, what the alternatives are and whether the distribution of risks and benefits is fair (e.g. Slovic, 2000). Who are the "experts" then? Several professions function as experts in relation to lay people, for example engineers, researchers and medical doctors. As described by Reventlow *et al.*, (2001), whereas general practitioners (GPs) conceive of risk as statistical risk calculated on a group level, patients perceive of risk through their own cultural lenses (Reventlow *et al.*, 2001). This creates a challenge when trying to communicate the risks of developing a particular disease and how it may be affected (or not affected) by a certain drug or lifestyle change (Reventlow *et al.*, 2001).

This can be seen in the context of medical education as well. For example, at the medical doctor's programme at Uppsala University, students are given a seminar assignment to convince vaccine sceptics to have their child vaccinated against measles, mumps and rubella (MMR). One group of students is given the role of the physicians and another group act as the parents. The students generally take the task seriously, but they usually use the numbers presented in the background material as a starting point. They try to convince the "parents" by pointing to the seriousness of the disease and the safety of the vaccine, referring to numbers and probabilities. Many students appear to question the rationality and morality of the sceptics. Do these parents not understand their own good? And why do they refrain from doing the best for their child? In light of the research done in the area of risk perception, it is not surprising that some parents refrain from having their children vaccinated. Furthermore, because this research shows how important values and emotions are when lay people deliberate on the acceptability of risk, it would be more effective to first find out what exactly the concerns are instead of assuming that they do not understand what is in their own interest. As we will see, it is also preferable from an ethical perspective.

Unfortunately, this attitude is not uncommon amongst professionals in different fields of expertise. Knowledge of how lay people perceive and deliberate about risks and benefits would help the experts attempting to make people understand the risks of a certain activity, technology or policy. It would contribute to the efficacy of risk communication. However, from an ethical perspective, it is more important to acknowledge that there are strong normative reasons to take the concerns of lay people into consideration. In this chapter, we turn our attention to the following question: How should risk communication be conducted in order for it to be ethically justified? In other words, what characterises ethically responsible risk communication?

In this chapter, a three-level framework is presented, with three corresponding conditions. Three examples are provided in order to discuss risk communication, and it becomes clear that although there are differences, there are also similarities between the communication of risks in the areas of health, technology and environment. Furthermore, there are differences and similarities between risk communication in the one-on-one context of general practice and in collective contexts such as public health or engineering.

Three conditions of responsible risk communication[1]

Risk communication has developed in three phases (Leiss and Powell, 2004). In the first phase, risk communication was seen as a form of education whereby the public should be informed about risk estimates. The underlying idea was that people in general did not understand science well enough and therefore needed to be educated. During the second phase, risk communication was transformed into a marketing practice with the aim of persuading people to adopt a certain message. In the current, third phase, risk communication is viewed as a participatory practice with which the gap between experts and lay people can be bridged (Leiss and Powell, 2004, p. 30). This is currently the ideal, supported by research. That does not necessarily mean that this is what happens in reality, and the examples discussed below illustrate the importance of this ideal as well as the lack of a participatory approach in some cases. Trust plays a crucial role both in bridging the gap between experts and lay people and in making risk communication effective (e.g. Persensky *et al.*, 2004). However, it should be emphasised that trust is not merely important for achieving effective risk communication. Arguably, trust requires trust*worthiness*, which in turn requires ethically responsible risk communication. Trust should be seen as the result of responsible risk communication, not merely as a tool for achieving effective risk communication. When a government, a powerful agency or a corporation aims to affect people's attitudes and behaviour in accordance with a certain goal, this activity entails moral responsibility. If this responsibility is taken seriously, this can create trust. It is important to see the difference between building trust in order to communicate effectively and building trust because one acknowledges one's power and takes responsibility accordingly. Thus, ethical risk communication is not merely required for instrumental reasons, but first and foremost for normative reasons.

In my view, there are three levels of risk communication where ethical analysis can and should be conducted: the message, the procedure and the effects (Nihlén Fahlquist and Roeser, 2015). Risk communication consists of a message concerning risk and a procedure through which the message is designed and communicated, which can be an iterative process. Needless to say, the risk communication process and the message communicated will have both intended and unintended effects. For example, if you send the message that smoking causes lung cancer or that obesity is unhealthy, people may or may not quit smoking and stop unhealthy eating habits; regardless, there may also be other effects, such as stigmatisation of smokers and obese people. Of course, risk communicators have some consequences in mind, for example, that people understand the dangers of smoking or overeating or that they continue to have their children vaccinated. However, it is also conceivable that the procedure and message will have unintended effects. Consequently, risk communication can be analysed at three levels: the procedure, the message and the effects.

These three levels provide for a conceptual model that can be used for ethical analysis. Moral values are relevant in relation to all of these stages. In common procedures for risk communication and stakeholder participation, emotions do not play an explicit role. However, given the research illustrating the importance of emotions and values in relation to risk, stakeholders should be explicitly encouraged to express their views and concerns concerning risks, as this will help to clarify moral concerns, thereby facilitating ethical reflection. In a similar vein, we see that messages about risk usually either consist of callous numbers or attempts to appeal to superficial gut reactions in a manipulative way. However, risk messages are inherently value-laden. These values can be expressed through moral emotions like care, empathy and respect. Such emotions are not primitive gut reactions but have cognitive content and a focus on moral values. This allows them to contribute to critical reflection (e.g. Roeser, 2018).

In sum, ethical issues in risk communication may arise in relation to the procedure, the message and the effects. Consequently, in order to ensure that values and concerns are adequately addressed in order for risk communication to be responsible, risk communication should fulfil three conditions. Ethically responsible risk communication requires a legitimate procedure, an ethically justified risk message and a concern for and evaluation of the effects of the message and procedure. We will now go through each of these conditions in more detail.

A responsible risk communication procedure

The first condition for responsible risk communication requires a legitimate procedure. The dominant view in social science literature is that a legitimate procedure for decision-making about risk should be a form of participatory technology assessment. Participation can take place at different stages of risk analysis, i.e. in risk assessment, risk management and risk communication. Many scholars argue that if relevant stakeholders are involved in political discussions and decisions about risk, the likelihood of achieving acceptance increases (Skitka *et al.*, 2003). People may accept a decision that they think has been reached fairly, even when they disagree with the actual outcome. However, it is also imaginable that a majority of people can be manipulated by a populist leader to adhere to ideas that are not based on science or rational thinking. A democratic procedure does not in itself guarantee an ethically justified message concerning risk. This is always a risk in a democratic society, noted in a more general fashion as early as Antiquity by, for example, Plato in *The Republic*.

It is worth emphasising that regardless of the efficacy of participatory approaches, it is ethically imperative in a democratic society that the public be included and allowed to participate in decision-making and communication about risk. However, including stakeholders in risk communication and regulation entails numerous challenges (cf. e.g. Bijker *et al.*, 2009; Lofstedt *et al.*, 2010). First, how should fairness and legitimacy be achieved within the participatory context? Second, who should be represented, by whom, and how should the representation be implemented? Jamieson and Briggs (2009), in discussing so-called stakeholder partnerships as a promising way to make risk communication more inclusive, argue that some of the challenges involve deciding how to achieve properly balanced and representative partnerships and how to make sure the representatives actually speak for their stakeholder groups (Jamieson and Briggs, 2009). It has been argued that even if risk communication becomes more inclusive and participatory, it is essentially a "relationship between unequal parties", and there will therefore always be an asymmetry in communicative initiative, informational privilege and risk influence between experts and lay people (Hayenhjelm, 2006). The party that initiates communication naturally has the privilege of setting the agenda and deciding with whom to communicate (Hayenhjelm, 2006). This will become clear when we explore the examples below. Healthcare staff and government agencies providing new parents with information and advice regarding infant nutrition and breastfeeding is an illuminating example of such an unequal relationship. Even with the best intentions, new parents are in a vulnerable position and want to do the right thing, which means that what and how they are told about risks to their new-born is a delicate matter. We will return to that issue later.

An additional problem with participatory risk communication is the fact that experts often disagree because contemporary risk issues are often ambiguous and value-laden and because experts can be biased due to conflicts of interest. However, sometimes they disagree because the evidence is overwhelming, such as with the health risks of smoking and the reality of anthropogenic climate change.

However, in some cases there is insufficient evidence, conflicting studies or results that show mere correlations rather than causation. Noticing the presence of expert disagreement in some cases does not imply a relativistic argument for abandoning the role of experts or leaving all risk decisions to lay people. As Munnichs argues, experts do have specific cognitive skills, knowledge and experience, and they enjoy public authority for this reason. However, their dominance may be questioned in an "adequate procedure of expert contestation" (Munnichs, 2004, p. 121). Such a procedure would be inspired by Popper, according to whom scientific claims should be exposed to a free competition of thought, where as many different expert views as possible are represented in a debate. Some experts are likely to share the same worries as the public, which means that those worries are represented in scientifically informed debates about risk. Although this may be part of the solution for making the procedure more legitimate, a morally responsible risk communication procedure also requires that the public actively participates in the debate. Lay people can add important perspectives because they have a broader conception of risk that comprises moral values (Slovic, 2000). To include moral values as well as emotions, has, as we have seen, also been emphasised by philosophers (Asveld and Roeser, 2009; Hansson, 2004; Roeser, 2007). This allows for a fair dialogue in which all of the affected parties and their emotions are heard and seriously considered.

Responsible risk communication: the risk message

As mentioned above, whereas a legitimate democratic procedure is important, it is not adequate. The second condition for ethically responsible risk communication requires that the risk message be formulated and presented in a way that is ethically justified. It has been illustrated by social scientific research on risk that different framings of risk messages have different results on people's perceptions and behaviour (Thaler and Sunstein 2008; Tversky and Kahneman, 1974). Since there are no neutral ways to design a message, it is crucial to analyse the moral values expressed by different framings. This entails two tasks. First, moral values expressed by risk messages should be made explicit (the formal task). Second, a debate about relevant moral values should generate a deliberated and morally reasonable risk message (the content-related task). As mentioned previously, a risk message usually contains or is based on scientific data, but it is also inherently value-laden. Risk messages usually concern the effects of human activity on health, safety and the environment, thereby touching upon important moral values such as well-being of humans, nature and future generations.

Furthermore, since it is impossible to communicate all risks, a selection must be made, and this selection is not a purely scientific issue (cf. Slovic, 2000), but also a normative one with ethical implications. How risks are presented or framed in messages has direct implications on how the audience is affected. Are any potential existing scientific uncertainties or controversies presented, and if so, how? As we will see below when discussing vaccination policy, the H1N1 vaccine that was provided in Sweden in 2009 was presented as completely safe and highly effective. It turned out later that it had not been tested to the necessary level of certainty and there were serious side effects.

One decision to be made is whether the message aims to convince people to behave in a specific way, or whether it should present various options for action with a list of pros and cons. Depending on the urgency, the former or latter option might be chosen. For example, in the case of a text message alert concerning an explosion accompanied by a life-threatening release of gas, the message might be simple and unambiguous: "close all windows due to dangerous gas" (cf. e.g. Jagtman, 2010). On the other hand, in the case of the long-term health effects of a certain activity, the pros and cons of alternatives may be presented in a balanced and nuanced way to leave more room for the decisions of recipients (Nihlén Fahlquist and Roeser, 2011; Roeser and Nihlén Fahlquist, 2013).

Responsible risk communication: assessing the effects

In addition to the procedure and the message, the effects of risk communication should also be analysed from an ethical perspective. Interestingly, evaluation of risk communication is under-theorised and seldom undertaken in practice (Fischoff, Brewer and Downes, 2011). However, three types of evaluation have been distinguished. First, there is the formative evaluation during the planning phase. Second, there is the process evaluation, evaluating the implementation of, for example, a campaign. Finally, an outcome evaluation analyses whether the goals of the campaign were met (Downs, 2011). Evaluating risk communication is important for reasons of effectiveness and efficiency. For example, if a public health authority has spent a substantial amount of money on an anti-smoking campaign, the outcome evaluation will ask whether the campaign *actually* led to a decrease in smoking (e.g. Downs, 2011). From an ethical perspective, it is important for this kind of evaluation to take the concerns of recipients or stakeholders into consideration by assessing whether they have been affected in an ethically acceptable way. In my view, a risk communication campaign can be both effective and ethically problematic. The H1N1 campaign in Sweden is, as we will see, such an example. Evaluating the unintended as well as the unintended effects of risk communication is important to ensure that all relevant values are taken into consideration. Furthermore, it is also necessary to evaluate whether the procedure and message are ethically justified and whether they address all relevant actors and values.

Hence, an outcome evaluation should assess whether the effects on the recipients are ethically legitimate. This in turn requires an attentive and sensitive attitude on the part of risk communicators (e.g. Nihlén Fahlquist and Roeser, 2015). We will now look at three examples and apply the three-level framework introduced above to see how risk communication should take these factors into consideration.

Example 1: nuclear energy[2]

Nuclear energy and risk communication procedure

Nuclear energy gives rise to intense debates and poses a great challenge for risk communication. The Fukushima accident in 2011 illustrates the need for a legitimate risk communication procedure. Nuclear risk communication in Japan prior to Fukushima appears to have been far from participatory and transparent. There were close ties between government and industry, and the approach to risk communication was top-down, leading to public distrust. Despite public concerns, nuclear plants were built in local communities, and when there were lawsuits, courts ruled in favour of government and industry (Figueroa, 2013).

Nuclear energy is a complex issue, technically as well as politically and ethically. The lack of a concise and consistent message in the case of nuclear energy illuminates the role of the process, that is, the way in which risks are communicated. Decisions concerning energy potentially have a great impact. The most important lesson from recent studies and arguments in social science and philosophy is that questions such as to what extent, if at all, we should use nuclear energy will not be answered by conventional methods such as risk analysis. There are moral values at stake, which means that decisions must be made in a democratic way, after serious debate about values and not merely about numbers.

A crucial question in relation to both procedure and risk messaging is how to make decisions about the so-called "decision horizon". The problem of an "indeterminate decision horizon" results from a lack of clarity concerning what the decision is about, for example when there is no agreement about the scope of the decision or about the problem it is supposed to solve. In cases of societal decision-making under uncertainty, a line must be drawn to demarcate the time horizon for the decision (Hansson, 1996). There is no one "right" horizon, but the horizon will vary depending on the perspectives of stakeholders. Different stakeholders are likely to have different views on what the risk decision is about, and hence about what to communicate. For example, nuclear energy producers might think that the relevant risk decisions are about acceptable risks in the context of already existing power plants. Greenpeace, on the other hand, might argue that the major decision to make is whether nuclear energy should be allowed at all. It should be acknowledged that the choice of a decision horizon and the fact that different stakeholders have different views about this is ethically relevant, because different decision horizons reflect different moral values of different stakeholders. For this reason, ethical risk communication requires a fair procedure in which all views are allowed in a discussion. This requires transparency and openness as well as respect and care for the public.

Nuclear energy and legitimate risk message

One possible pitfall is that a procedure taking all possible viewpoints into consideration might result in a lack of leadership and a "risk information vacuum" if no decisions are reached and the discussion continues (Leiss and Powell, 2004). A lack of message is thus also a message, and it is not only ineffective, but in some cases unethical, to refrain from communicating risks. From an ethical perspective, a risk information vacuum is disrespectful and shows a lack of care and concern for other people. Furthermore, the procedure and the message are intertwined and shape each other. Hence, the second condition requires that risk messages concerning nuclear energy be ethically deliberated and moral emotions clarified.

Nuclear energy entails risks at different levels. First, there are safety risks relating to accidents and the immediate repercussions in terms of people's health. The potential harmful events include primarily diffusion of radioactive substances. Second, the long-term environmental risks are due to the uncertainties concerning the long-term effects of nuclear waste management. The former risks are more short-term and can affect people living now, such as in the case of Fukushima and Chernobyl. The latter are more long-term and affect future generations. Risks can be due to technological failures, human error or destructive intentions (e.g. terrorists). Materialised risks can affect human beings and the environment, now and in the future. Matters are even more complicated because there is a lot of uncertainty concerning the risk of production and use of nuclear energy (Taebi, 2011).

Regardless of whether the message concerning nuclear energy is that "It is safe" or "It is not safe", values are expressed, because the messages are based both on probability calculations and on views about moral acceptability. As has been argued by social scientists and philosophers, quantitative information is not sufficient to decide whether a risk is acceptable or not (Krimsky and Golding, 1992; Shrader-Frechette, 1991). For example, a certain activity is always only safe or risky in relation to something else, and what we choose to compare it to is a normative choice. Furthermore, the very concept of "safety" suggests that the risks are acceptable, but acceptability is a normative notion and requires the normative reflection of empirical data. Safety is, in this sense, a thick concept, that is to say, it is both descriptive and normative at the same time (Möller, 2012). After the Fukushima accident, several countries revised their official risk messages concerning nuclear energy. In Germany, the previous message that nuclear energy was safe enough was replaced by a message stating that nuclear power is not safe enough and should be banned. Within a few days after the new message was announced, German Chancellor Angela Merkel decided to immediately shut down nine reactors (Ionescu, 2012). Messages such as these concerning the risks and safety of nuclear energy contain values that should be made explicit and discussed when deliberating and deciding what the message ought to be.

The second requirement for ethical risk communication points to the importance of a deliberated and justified risk message. Borrowing a concept used in the ethics of technology, we could say that a risk message should be designed in a "value-sensitive" way in order to be ethical (See e.g. Flanagan, Howe and Nissenbaum, 2005; Friedman, 2004; Manders-Huits, 2011; Van den Hoven, 2005). Applied to nuclear energy, values in risk messages concerning nuclear energy should be made explicit and should be deliberated. A complex range of issues can be included in messages that all reflect values. For example, stating that nuclear energy is necessary in order to mitigate climate change may express values concerning the need to reduce anthropogenic CO_2 emissions, but could also be seen as expressing a lack of concern for the ability of future generations to deal with nuclear waste. Risk messages must be ethically scrutinised and deliberated and values must be explicated.

Values in risk messages can be made more explicit by appealing to moral emotions such as concern about our grandchildren's future. Fear and feelings of responsibility, for example, could help us understand why it is wrong to expose our grandchildren to the risks of climate change or the risks of nuclear waste (Nihlén Fahlquist and Roeser, 2015; Roeser, 2012a, 2012b). Of course, this is challenging, since emotions can be abused in communication, for example, by appealing to instinctive responses in order to manipulate people to engage in certain behaviour, as marketing experts are very well aware of (cf. e.g. Buck and Davis, 2010). However, by appealing to potentially reflective emotions such as sympathy, compassion and feelings of responsibility, risk communication can stimulate critical and ethical thinking (Roeser, 2012a, 2012b). When communicating about values involved in nuclear energy, one could include pictures and stories of victims of climate change and coal mining on the one hand and victims of nuclear disasters and scenarios of future generations dealing with nuclear waste on the other hand. One could also include visualisations of our ecological footprint and energy consumption to increase awareness of our wasteful lifestyle. Additionally, one could include pictures of more or less successful installations of alternative energy sources such as windmills, solar panels and water dams. By showing the multifaceted nature of energy policy, one could increase awareness for the complexity of this issue. Such a risk message could communicate that nuclear energy might not be an all-or-nothing decision but rather a possible ingredient in a future energy mix about which we all must make up our minds and in which we all have a responsibility. Such a risk message would explicate the values involved, thereby facilitating sound and moral decision-making concerning messages about nuclear risks (Nihlén Fahlquist and Roeser, 2015).

Nuclear energy and the effects of risk communication

The third condition for ethical risk communication requires an ethical evaluation of the effects of risk communication. If we consider risk communication concerning nuclear energy in Japan, there a number of ethically problematic factors from before, during and after the accident, which led to a (justified) lack of trust. According to some studies, the Japanese government did not reveal key aspects of the accident and its associated risks. The government understated the risks and wrongly sent the message that nuclear power is safe (Figouera, 2013; Shrader-Frechette, 2011). The government's attempts to convince the public that it was in control backfired, since events revealed that the government was not in control (Figueroa, 2013). Thus, the ethically flawed risk communication procedure and message led to anger, frustration and distrust. Arguably, these are justified moral emotions in this case.

Since the Fukushima accident occurred after information became accessible on the Internet in real time, it is clear that electronic communications do not ensure adequate risk communication. However, experts did not themselves have access to all information, resulting in confusing and contradictory messages. Interestingly, in Germany a risk communication effort in Stuttgart following the Fukushima accident focused on direct meetings and discussions with the public as an alternative to media representations (Ionescu, 2012). On 24 March 2010, the Institute of Nuclear Technology (IKE) invited experts and lay people to the University of Stuttgart to participate in a discussion of the technology and safety systems of the reactors and the causes of their failure. People could express concerns and experts could provide and explain the information to which they had access. According to Ionescu, this event was very successful as it "filled a void of trust and expertise" caused by confusing media coverage. The success of the discussion shows that even in situations of crisis there is a need for more public engagement (Ionescu, 2012). In this case, the chosen procedure had positive effects in terms of trust, in contrast with the situation in Japan.

This illustrates that different risk communication strategies can have different results in terms of effectiveness as much as in terms of trust and moral responsibility. This could also provide invaluable information for future decisions on messages and procedures concerning risk. Assessing the effects of risk communication should be an iterative process in which the message and procedure can be reassessed and re-evaluated in the light of the possible effects. People could be asked how they experienced a certain risk communication campaign, how they were affected and why. This should also involve asking them about ethical concerns, about whether people experienced the campaign as helpfully providing further insights and whether the risk communication has made them feel uncomfortable and if so, how. There is important information to be gained if we understand why risk communication made people feel uncomfortable. Was it because it made them aware of their own responsibilities, or was it because it provoked a sense of hopelessness, making them passive? As Hulme (2009) has pointed out, this is a possible pitfall in communication about climate change, and we can extend this to communication about nuclear energy. From an ethical point of view, it would be desirable to practice forms of risk communication that encourage people to be actively engaged, to see a possible role for a contribution of one's own rather than seeing oneself as a victim of inevitable circumstances or higher powers that are ignorant or dismissive of people's views. This could potentially increase awareness about climate change and critical reflection about responsible energy mixes and a possible role for nuclear energy. If the effect of a certain risk communication effort is emotional reflection and a willingness to take responsibility for energy-related issues, that is a desirable effect. If the outcome is a lack of trust or a sense of hopelessness and passivity, arguably, risk communication should be reconsidered and radically reformed (Nihlén Fahlquist and Roeser, 2015).

Example 2: vaccination policy[3]

During the last couple of years, measles has reoccurred in Europe. In March 2017, 11 cases were reported in Sweden. Healthcare staff and government agencies remind people that it is highly important to have their children vaccinated. It is not uncommon to hear physicians or government representatives state that it is wrong not to vaccinate your children (Larsson, 2015). The EU Commissioner for Health and Food Safety recently stated that public distrust in immunisation is a serious threat to public health (European Commission, 2017). Even though vaccines are promoted by physicians and agencies, vaccine resistance, or vaccine scepticism or refusal, has been increasing (Dubé *et al.*, 2013; Hough-Telford *et al.*, 2016). Parents refusing or hesitating to have their children vaccinated have been described as a major public health problem in many countries where both individual and community risk of vaccine-preventable diseases is increasing (Smith *et al.*, 2011). The growing hesitancy or refusal to vaccinate children has led to a situation where the WHO goals to eliminate measles is now challenged. The vaccine for measles, which is part of the MMR (measles, mumps and rubella) vaccination, is effective and safe, and the side effects are minor. Experts are worried that the rate and level of vaccination will continue to decrease. Against this background, it has been argued that vaccination should be mandatory (Gustavsson, 2015). Alternatively, it has been suggested that vaccination should be required in order for people to get access to other public goods. For example, in Stockholm, it has been suggested that children should possibly not be allowed to start preschool unless they are vaccinated (Edholm and Starbrink, 2017).

Different countries adopt different strategies in relation to the problem of vaccination and the challenge of achieving and maintain protection and herd immunity. However, mass vaccinations, even if they are not mandatory but merely strongly recommended, entail ethical problems that must be thoroughly discussed in society and managed responsibly by the government and its representatives. Ethical aspects of vaccination are a recurrent theme in public health debates (Dawson, 2009; Sorell, 2009) However, the most common arguments focus on the moral acceptability of refraining from having one's children vaccinated, the ethical acceptability and necessity of mandatory vaccination and the conflict of values involved in vaccination as a public health measure. The central ethical dilemma of public health is how to promote the health of the population while respecting individual liberty and autonomy. In this chapter, it is assumed that vaccination, as public health and health promotion generally, within certain limits, is ethically justified. However, this does not mean that vaccination should be promoted at all cost and in any way. Instead, there are good reasons to discuss the ethical boundaries of risk communication concerning vaccination and to argue that the government needs to take responsibility for the individuals affected by side effects. Thus, at this point, the most urgent ethical question is not whether mass vaccination is right or wrong, or whether it is ethically acceptable to refuse vaccination. Instead, we should discuss *how* vaccination should be communicated about in order to be ethically justified. Furthermore, given that vaccination should be promoted, what responsibility should governments take for those individuals who inevitably will be affected negatively by (particularly new) vaccines? These questions are crucial for two reasons. First, an ethically well-deliberated vaccination policy is normatively important in a society based on notions of fairness, democracy and respect for individuals. Second it is more likely to contribute to increased trust and vaccine coverage. It has been acknowledged that taking ethical considerations seriously may be crucial to facilitate global vaccination (Schwartz and Caplan, 2011). Although some people will continue to refuse vaccination, there is currently a group of "vaccine hesitant" people, that is, people who could possibly change their mind (Dubé *et al.*, 2016). This group of people might be convinced if the risks and benefits of vaccination and disease are discussed in a respectful and ethically deliberated way.

Thus, two arguments concerning responsible risk communication in the case of mass vaccination and vaccination hesitance are introduced in this chapter. First, the procedure of risk communication should be scrutinised, and risk communicators, such as healthcare providers and government agencies, should be attentive to the concerns of lay people. Second, the effects of previous vaccination policies should be analysed in order to improve future risk communication. In particular, I argue that the government should take responsibility for individuals suffering from side effects, particularly of new, less-tested vaccines. The chapter focuses on the Swedish context as it links the current scepticism to the H1N1 vaccination programme in 2009, but it is relevant to all countries reflecting on how to manage vaccination responsibly.

Measles and vaccination

Measles is one of the most infectious diseases and one of the leading causes of global childhood mortality (WHO, 2017). In 2016, measles outbreaks were seen in several EU/EEA countries. In 2017, the number of cases continues to increase (ECDC, 2017). In 2015, 134,200 people died from measles, which equates to 367 deaths every day (WHO, 2017).

The Swedish vaccination programme, offered free of charge for all children, includes vaccination against nine diseases, including measles. As a result, and also due to the traditionally high level of trust in authorities and medical science in Sweden, there have been very few incidences reported yearly. However, in March 2017, 11 people were reported as infected, and in some regions the vaccination coverage is relatively low. This has led to a discussion about the need to vaccinate children, as well as adults, and there is growing concern that vaccination rates will decrease due to scepticism and unjustified fear of side effects.

What to do about the "anti-vaccinationists" and other sceptics?

Vaccine sceptics have participated in public debate since the smallpox vaccination was introduced in the nineteenth century. The Vaccination Act was passed in the United Kingdom in 1853. It made smallpox vaccination mandatory for infants in the first three months of life, and parents were made liable for fines or imprisonment in cases of non-compliance. Vaccine resistance began almost immediately. According to some studies, the arguments used by sceptics have remained more or less the same ever since (Wolfe and Sharpe, 2002). However, developments in traditional and new media have naturally changed the extent and method of communication (Saint-Victor and Omer, 2013).

Although vaccines are improving, there is a growing decline of confidence in them (Larson *et al.*, 2011). This has led to a growing concern amongst experts that this lack of trust in vaccines will lead to less protection, more frequent disease outbreaks and threats to herd immunity. As argued by Blume, it may be appealing to see sceptics as a united movement because it "unites health professionals behind a banner of reason" (Blume, 2006, p. 639). The sceptics are frequently seen as less educated, anti-science, selfish, irrational and driven by fear and anxiety (Blume, 2006).

For example, in Sweden, the famous paediatrician Hugo Lagercrantz states in a newspaper article that parents who refrain from having their children vaccinated risk harming both their children and others. It is, he claims, a disloyalty to fellow human beings (Larsson, 2015). However, the focus on organised irrational or science-hostile opponents diverts attention from any valid concerns that parents may have (Blume, 2006). Studies show that refusal, scepticism and lack of trust could be based on social, political or safety-related assumptions and concerns, which indicates that it is a complex issue (Larson *et al.*, 2011). Furthermore, although some concerns are perennial, others are new, and each vaccine is introduced in a different historic context. When examined carefully, it may turn out that although some concerns are unfounded, others may be more valid. There is a contemporary discussion about "alternative facts" and an unscientific and relativistic notion of truth that is highly problematic. If this is what lies behind vaccine refusal, there are good reasons to correct misunderstandings and educate people about scientific methods. However, the public's concerns are more complex and nuanced and affected not merely by distrust of science, but driven by a mix of scientific, psychological, socio-cultural and political reasons (Blume, 2006; Larson, 2011). Decision-makers need to understand this mix and the source of people's concerns in order to achieve communication based on trust and to increase vaccine rates (Larson *et al.*, 2011). These insights have led to a new concept being used in addition to vaccine refusal, namely "vaccine hesitancy" (Dubé *et al.*, 2013; Dubé *et al.*, 2016). The concept was introduced in order to show that there can be many different reasons and ways of being hesitant about some or all vaccines.

Research in risk communication has identified three factors that affect the extent to which an individual or institution is trusted: perceptions of knowledge and expertise, openness and honesty, and concern and care (Slovic, 2000). To frame the debate as one between knowledgeable experts and uneducated or unreasonable opponents does not take these insights into consideration, which may be one reason why risk communication efforts do not appear to increase vaccination rates.

Taking concerns seriously

The ethical discourse about vaccination is often set up in terms of the utilitarian goals of public health and opponents' claims relating to individual liberty and the right to autonomy, that is to say, as a conflict between different values. However, the ethics of vaccination is not only a matter of potentially conflicting values and choices that are either right or wrong. The way experts communicate risks and benefits in the context of vaccination matters as well. As we have seen, social scientists have shown that perceived knowledge, openness/honesty and concern/care affect public trust. Trust is sometimes conceived of as a way to achieve higher rates of vaccination. It is important to keep in mind that trust is not merely a means, but ideally the result of good and ethically justified public health activities.

To treat the public and their concerns as overly emotional and, therefore, irrational can be seen as part of a tradition of thought that dichotomises Reason and Emotion, a dichotomy that has been rightly criticised, as we have seen. Again, it is important to see emotions as a starting point of a discussion about the moral acceptability of risks.

These insights, whose relevance in the context of nuclear energy we have seen, should be used in the design of risk policy and communication about vaccination. As Larson *et al.* argue, although extremist anti-vaccination groups of people who are unable or unwilling to change their minds exist, many people *could* change their mind. The focus should be on this group and on building and sustaining trust with them.

Against this background, the *method* for communicating about vaccines and their risks should be brought to the fore. It is not enough, from an ethical perspective, to inform the public of the scientific evidence and probabilities concerning risks and benefits. As I have argued, it should be done in an ethically acceptable way, taking lay people's ethical values seriously (Ruddick, 1989). Even though vaccination is a question of public health and scientific evidence, it is also a question of ethical norms and values. Science can provide evidence that vaccines are safe and likely to contribute to a reduction or elimination of certain diseases. However, whether everyone has a responsibility to have their children vaccinated, or whether it is a duty to be vaccinated, is a moral question. Just like in the case of nuclear energy, lay people should take part in the ethical discussion.

As argued by Roeser and Pesch, in relation to moral norms and values, "all citizens are *prima facie* on equal footing" (Roeser and Pesch, 2016).

Vaccination policy – message and procedure

Let us look at the three levels again, this time in relation to vaccination policy. The question then is the following: *How* should moral emotions and values be used in the discussion on vaccines?

Public health activities are "purposeful attempts to influence target audiences' perceptions and behaviours" through persuasive interventions and social marketing (Noddings, 2002). It is, therefore, an inherently value-laden activity.

In the context of vaccination policy and risk messaging, vaccination can be framed in different ways. It could be stated that everyone should be vaccinated and have their children vaccinated in order to protect the individual or one's family. Alternatively, it could be promoted as a collective responsibility for vulnerable groups or as a duty for citizens of liberal European welfare states to contribute to population health. Messages linked to personal responsibility and culpability are common in public health (Guttman and Salmon, 2004). In order to make decisions about the framing of risk messages, we need to have a respectful discussion in society including all views, which illuminates the procedural aspects.

In addition to examining the message ethically, it is important to validate the procedure by which information is provided and how decisions concerning new vaccines are made. In a democratic society, where the values of openness, fairness and inclusiveness are embraced, it is important to promote transparency of expert deliberations because, as argued by Verweij, it is not reasonable to simply expect compliance in all new vaccination policies (Verweij, 2013). This might seriously compromise existing trust, even in countries where vaccine coverage is currently satisfactory.

As discussed in the introduction to this chapter, recent research in social science highlights the need to include all stakeholders in order to achieve trust and a legitimate process. It is reasonable, in a democratic society, to expect a respectful two-way dialogue instead of a one-way transmission of information from government authorities and healthcare staff to recipients such as lay people. If relevant stakeholders are involved in political discussions and decisions about risk, it increases the likelihood of achieving acceptance and maintaining trust (Schwartz and Caplan, 2011; Skitka *et al.*, 2003).

Just as in the context of nuclear energy and technological risks, moderators of public debates could ask lay people questions about their fears, concerns and possible solutions, which would promote dialogue, respect and understanding of different perspectives (Nihlén Fahlquist and Roeser, 2015).

Effects on trust – the H1N1 vaccination policy in Sweden

We have discussed ethical aspects of the message and procedure for risk communication in the context of vaccination policy. The third condition for ethical risk communication concerns the effects of risk communication campaigns and policies. The effects of vaccination policy and communication deserves a more extensive discussion, especially in the Swedish context. Trust in experts and policy is likely to partly be a consequence of how experts communicated risks and benefits of previous policies. In the Swedish context, it is important to discuss the way in which the H1N1 vaccination policy in 2009 and its effects were managed.

In 2009, the WHO declared the H1N1 influenza a pandemic. The European Medicines Agency (EMA) promptly authorised three vaccines. In Sweden, the government implemented a vaccination programme with the aim of offering free vaccination for all inhabitants using the Pandemrix vaccine. Everyone was encouraged to get vaccinated, both for the sake of their own health and to prevent others, particularly vulnerable groups, from getting sick. The message communicated was that it is everyone's responsibility and the only right thing to do. The vaccine was safe, according to government agencies, with no long-term or serious side effects. After the threat of H1N1 faded, a larger than usual group of adolescents were diagnosed with narcolepsy. Studies have now shown that there is a correlation between the vaccine and narcolepsy. According to a study by the Swedish Medical Products Agency (MPA), children under the age of 20 were three times more likely to get narcolepsy after having received the Pandemrix vaccine (MPA, 2013). Without going into the details of this complicated case or attempting to evaluate how the Swedish government managed this situation, the case is illustrative of the reasonable requirement that risk communication should take the effects of the communication on recipients into account, and that the responsibility of risk communicators should continue after a crisis. Most likely, the parents whose children were diagnosed with narcolepsy experienced anxiety, guilt, fear and anger after the campaign. They were initially told that it is highly important that everyone, including their children, receive the vaccine in order to protect themselves and others and that there were no serious side effects of the vaccine. Still, after the fact, they were left to take care of their almost grown-up narcoleptic children who will never be able to live normal lives, have normal jobs, etc. There should be a special responsibility to follow up the consequences, to listen to individuals suffering from side effects and to attend to their new needs.

Unless the issues arising in the context of the H1N1 policy are dealt with in a satisfactory way, in the eyes of the public, experts and governmental agencies engaging in health risk communication are likely to receive diminishing trust and compliance. It would not be surprising if this affected not only influenza vaccination or new vaccinations, but also the relatively safe and evidence-based measles vaccination.

Of course, the media has substantial power to influence people's beliefs about vaccines. For example, in 2015, concerns about the HPV vaccine in Denmark arose as a reaction to a television documentary entitled "The Vaccinated Girls – Sick and Betrayed". It is reasonable to expect the media to be self-critical as well and to take responsibility for the communication of risks and benefits, as well as science and values.

From a scientific and public health perspective, it is quite obvious that the MMR vaccination, despite certain risks, is a good thing and should be promoted. However, this does not imply that vaccine sceptics should be seen as the enemy or that the government and its representatives should continue to state that the only right thing to do is have ones' children vaccinated and that any concern is invalid.

Taking responsibility for individuals affected by side effects

The intentions of vaccinations, as with all public health policy, are good. The idea is to improve people's lives and health status and to use collective resources in an efficient manner. However, even though intentions are good, the effects, or some of the effects, may be harmful or otherwise ethically problematic, which is why ethical analysis of potential effects is necessary.

It is highly likely that certain side effects of vaccines must be accepted, and that mass vaccinations are inevitable and ethically justified. However, it is important to be aware that the basis for this kind of policy is collectivist, and it is likely that there will be individuals affected negatively by this kind of policy. The government should acknowledge this and be ready to take responsibility for such individuals affected negatively by a well-intended vaccination policy. Unless this is done, it is hard to see how trust can be restored.

One way of taking responsibility is economic compensation. When a likely causal link was established between Pandemrix and narcolepsy, some people who developed narcolepsy after being vaccinated were economically compensated through the Swedish Pharmaceutical Insurance. This insurance is financed by the shareholders of a corporation called LFF Service AB (the Swedish Pharmaceutical Insurance Service), which is comprised mainly of pharmaceutical companies (LFF, 2017). This method of compensation was criticised because there were several financial and age-related restrictions concerning eligibility for compensation. For this reason, the government and opposition parties made a joint decision in 2016 to have the government compensate those affected. The insurance is still primarily used for compensation, but the government supplements it. To be eligible, the age restriction from the previous compensation scheme was lifted. However, a maximum compensation sum of SEK 10 million was set. According to the association for narcoleptics, this sum was far from satisfactory and will not cover the costs for young people who will not be able to have a career. In Norway, the amount of compensation depends on how seriously affected the person is, and in Finland, patient insurance can be supplemented by the government based on the needs of the affected person (Svensson, 2016). In order to restore trust, it is highly important to listen to the arguments of the stakeholders involved in the H1N1 case, such as the association for narcoleptics concerning compensation. It could also be a good idea to compare the policies adopted by different countries in order to come to a conclusion regarding a given policy's ethical acceptability. Arguably, all of these tasks, including listening to stakeholders and comparing with other countries, are required by whomever is responsible for those individuals who are negatively affected by public health policy. Having said that, economic compensation is not equivalent to taking responsibility. Most importantly, ethical reflection and discussion is needed in order to decide what it means to take responsibility for these individuals.

Even if there are good reasons to promote mass vaccination as an ethically justifiable policy goal, there are still important ethical aspects to discuss, especially since public health communication is inherently value-laden. As a society, we probably have to accept some side effects, such as individuals affected negatively by public health policies, but this makes it crucial to ensure that individuals who are harmed are taken care of. This is important for normative ethical reasons, but also in order to restore and maintain trust in experts and vaccines. What exactly this entails in different cases and contexts should also be discussed in public.

Example 3: infant feeding recommendations[4]

We will now turn our attention to a third and final example that illustrates the importance of taking ethical aspects into account in risk communication. This example is similar to the case of vaccination in that it is a public health case that is also characterised by good intentions but that nevertheless causes harm to a group of people: breastfeeding recommendations.

The breastfeeding message

There are two main ways to feed an infant: breastfeeding and bottle feeding with substitute formulas. The information about infant feeding provided to parents strongly favours breastfeeding above formula feeding. "Breast is Best" is the slogan that is used to promote breastfeeding in the USA and the UK. The official Swedish policy is that although formula can be given if there is a problem, breastfeeding is the best option for babies. Breastfeeding is described as economical, environmentally friendly and cost-effective (Stockholms Läns Landsting, 2017). Packages of formula in the Netherlands and Sweden carry a message that "giving your milk to your baby makes a big difference to both your baby's health and yours.

And every day counts: the longer you feed your baby mum's milk, the more they benefit". In the UK, mothers are told by the National Health Service that "breastfeeding gives babies the best start in life" (UNICEF/NHS, 2018). The WHO and UNICEF promote breastfeeding and declare that "all women should be able to practice exclusive breastfeeding and all infants should be fed exclusively on breast milk from birth to 4–6 months of age" (WHO/UNICEF, 1990). Manufacturers of substitute formulas are not allowed to market their products (WHO, 1981), one consequence of which is that, for example, in Sweden, they do not even provide information on how to prepare formula milk on their websites.

According to most sources that provide information on infant feeding, breastfeeding is the superior way to feed an infant. According to these sources, breastfeeding is the only natural way to feed a baby, facilitates bonding and a close relationship between mother and baby and is more convenient and less expensive than bottle feeding. Breastfeeding is an easy way to help a mother get back in shape after pregnancy. Breastfeeding allegedly guarantees the right amount of nutrition at the right moment, as breast milk adjusts its makeup during the development of the baby. Formula milk cannot provide all the nutrients in breast milk since they are too costly to produce artificially. Breastfeeding was previously said to prevent atopic dermatitis, allergies and asthma. That link was subsequently questioned and is no longer stated as true. Breastfeeding is supposed to strengthen the immune system. Breastfeeding is sometimes said to reduce the risk of breast cancer and ovarian cancer. Mothers who breastfeed will allegedly have more intelligent children than mothers who bottle-feed. Breastfeeding is said to prevent obesity and diabetes at a later age (Moreno *et al.*, 2011). Given this extensive list of alleged advantages of breastfeeding that can be found in many official sources on breastfeeding, it is not surprising that many mothers want to breastfeed their children. They are also strongly encouraged to do so by midwives, hospitals and doctors. The WHO offers a "model chapter" for textbooks intended for health professionals and medical students, in which it is strongly suggested that mothers are encouraged to breastfeed (WHO, 2009). WHO/UNICEF has set up a "Baby Friendly Hospital Initiative" (BFHI) that is implemented in many countries and provides hospitals with a certificate if they promote breastfeeding. Governments work together with the breastfeeding-lobby organisation La Leche League (LLL) in order to promote breastfeeding (for example, for information on the Dutch policy, cf. Lanting and Van Wouwe, 2007b, p. 5). However, in practice, breastfeeding does not always work out so smoothly.

"All mothers can breastfeed"

A lot of sources of information on breastfeeding emphasise that all mothers should breastfeed. Indeed, many women have no problems with breastfeeding and enjoy it. However, there are also many women who experience physical problems while breastfeeding. A number of women actually find breastfeeding to be more painful than giving birth (Bick *et al.*, 1998; Cooke *et al.*, 2007). In fact, even though most women plan to breastfeed for a substantial period and think it is the best food for infants, many new mothers experience physical challenges at least in the beginning, and some struggle for the whole breastfeeding period (Kelleher, 2006). Many mothers give up breastfeeding. For example, in the Netherlands, although 81 per cent of the women breastfeed exclusively after birth, only 48 per cent do so after one month, 30 per cent after three months, and 23 per cent for at least the recommended six months (Lanting and Van Wouwe, 2007a).

The most important reasons mothers stated for giving up breastfeeding within or after the first month was that it was painful and that they did not have sufficient milk. Other commonly mentioned reasons for stopping were: concerning the mother: tiredness, weight loss, use of medication and better sleep at night; concerning the baby: tummy aches and restlessness, dissatisfaction, hunger and low body weight. Some women mention work as a reason to stop breastfeeding (for all this, cf. Lanting and Van Wouwe, 2007b, p. 16). Kukla describes how some women who have been sexually abused may experience problems with breastfeeding because it can trigger abuse memories, and women with eating disorders who are not comfortable with their own bodies may not feel comfortable breastfeeding (Kukla, 2006). In another study, several women said that breastfeeding gave them a feeling of loss of self and agency. One woman described how breastfeeding made her experience a sense of alienation between body and self, a feeling that was exacerbated because she was also experiencing physical pain (Schmied and Lupton, 2001). Some mothers are unable to breastfeed their babies at all due to varying circumstances such as a difficult or early birth, inverted or extremely sensitive nipples, infections or limited milk production. However, women in these situations often try their hardest to breastfeed, sometimes beyond exhaustion. Hospitals and midwives provide assistance with breastfeeding; there are even lactation experts. Despite this assistance, many women give up breastfeeding due to problems. The attitude of nurses and midwifes varies between support for the particular circumstances of the mother and baby to an attitude of telling mothers to keep trying, which is often experienced as paternalistic and aggravating an already stressful and emotional situation, especially for women who would like to breastfeed but simply do not succeed (Wolf, 2007). Trying to breastfeed pushes these mothers to and over their limits, which, apart from risking the mother's health and well-being, also risks negative effects on the mother–baby relationship. Although problems with breastfeeding happen frequently, this is hardly ever mentioned in official documents, giving women the impression that they are alone with their difficulties. Rather, it is generally emphasised that "all women can breastfeed" – implying "if they want". Stating that all women can breastfeed and combining it with statements like the following: "Every day you breast feed makes a difference to your baby's health now and in the future" (UNICEF/NHS, 2018) are likely to have negative effects on non-breastfeeding mothers. The implication for these women is that they are not trying hard enough and that they are not sufficiently concerned with their child's current and future health. The official information on infant feeding focuses almost exclusively on the advantages of breastfeeding, seeming to imply that bottle feeding constitutes a risk. For example, in the WHO chapter intended for health professionals and medical students, the following is a suggestion about how to explain this to new mothers: "Explain about exclusive breastfeeding, remind the mother that she does not need to give anything else before the baby is 6 months old, and that feeding bottles are dangerous" (WHO, 2009; Nihlén Fahlquist and Roeser 2011).

The naturalness argument

The current message supported and promoted by WHO and national and local bodies contains a mixture of scientifically supported, evidence-based statements, as well as semi-scientific and value-laden statements. To say that breastfeeding protects the infant and possibly the mother against diseases are statements of fact that can be supported and falsified by research. Statements regarding convenience, however, are very subjective, and statements concerning the naturalness of breastfeeding are highly value-laden. A common fallacy in thinking about risks is to assume that nature is benign, and technology is malicious (Slovic, 2000). However, this view of nature is empirically false, as there are lots of dangerous substances and events in nature. Yet, we see that an argument that is often used in favour of breastfeeding is that it is more natural (Rikshandboken, 2018). Naturalness is usually considered and presented as a good thing, a value to be upheld and an ideal that we ought to strive for. Consequently, it is not surprising that parents want to feed their babies in a "natural" way. However, in spite of the positive connotations of naturalness, there are many philosophical problems with the concept, which is vague, value-laden and even political (Soper, 1995). First, it is extremely difficult to say for sure what is natural and what is unnatural. Second, many things we appreciate today would not be considered natural, yet we prefer the new unnatural instead of the traditional, natural thing. For example, people today see glasses, central heating, means of transportation and information technology as necessary tools that facilitate daily life tremendously. There are also numerous examples of dangerous but purely natural things, such as toxic substances (cf. Hansson, 2003). Consequently, "natural" does not necessarily mean that something is beneficial to us, or equivalent to good or safe.

Norms concerning motherhood

Breastfeeding has been described as part of an "ideology of total motherhood". The latter, Wolf argues, is

> a moral code in which mothers are exhorted to optimize every dimension of children's lives, beginning in the womb, and its practice is frequently cast as a trade-off between what mothers might like and what babies and children must have.
>
> (2007, p. 615)

As Kukla argues, there are many norms that "good mothers" are supposed to live up to (Kukla, 2008). Otherwise, they are punished and shamed. Mothers are treated as "risk managers", as if their main responsibility is to assess and manage risks to which their child is exposed. Reproductive warnings, which are proliferate in contemporary society, single out women as uniquely responsible to "make choices that minimize risk for others".

Assumption of maternal superiority

In addition to the value-laden message that breastfeeding is a better option because it is supposedly more natural, the current message is based on both a heteronormative assumption and an assumption of good motherhood. Fathers and other possible caretakers are not even mentioned. This illustrates how the assumption of maternal superiority underlies the debate on infant feeding (e.g. Law, 2000). The current focus on breastfeeding as superior to formula feeding disregards the role of fathers, adoptive parents, gay couples and other non-breastfeeding families. Most babies have a father or a second parent, and if the mother breastfeeds for at least six months – and sometimes for years, which is also encouraged by official sources – the relationship between the father/other parent and the child is likely to suffer. The baby will form a strong bond with the mother, but there is a risk that this will happen at the expense of the other parent.

If there is a second parent, it is presumably a good thing for the child to bond with this other parent as well. Furthermore, the second parent arguably has a right to bond with his/her child, and this should not be neglected. For traditional and conventional reasons, most societies view mothers and fathers differently. Breastfeeding is thus not seen merely as a matter of feeding and bonding with one's baby, but as deeply connected to being a mother and even to the role of women in society. However, the arguments for seeing mothers as primary caretakers can be contested, to say the least. Men can be equally good parents, and fathers should arguably be encouraged to take a more active role in parenting their children. There is a lot of discussion about how to get fathers more involved, but the norms of exclusive motherhood and complete breastfeeding may counteract the inclusion of fathers in family life (Blum, 1999). The idea that men can be just as good parents as women even when the child is still an infant means that they can be equally important as providers of infant food. This allows for a more equal distribution of tasks between parents. In addition, if fathers are given the opportunity to participate in the feeding of infants, they are able to share in early emotional bonding with their child. Bonding with one's two loving and caring parents should not be neglected to the benefit of exclusive bonding with the mother.

Feminists have been engaged with the issue of breastfeeding for some time, and the relationship between feminism and breastfeeding is complex, not least because there is not one feminist theory, but several branches. Difference feminists want to upgrade women's traditional work and women's experiences and question the male norms on which society is based. Difference feminists are often pro-breastfeeding because it is considered a female experience. The current social and institutional structures that prevent women from breastfeeding are criticised. Against this background, it has been argued that breastfeeding is a right and that society should facilitate breastfeeding (Stuebe, 2009; Van Esterik, 1994, 2011). In contrast, liberal feminists emphasising equal treatment and women's inclusion in the paid labour market have been sceptical. For example, the American National Organization for Women (NOW) was divided concerning the recommendation in 1997 by the American Academy of Pediatrics (AAP) to breastfeed for 12 months. Initially, a spokesperson for NOW called it anti-working women, but later the vice president said that breastfeeding is a personal choice, and that business and society should make sure working women are able to breastfeed if they choose to do so. The background of this debate lies in the discussion between feminists arguing that women need protective laws and special treatment and those who argue that gender equality requires that women be treated just like men without special treatment.

Thus, the message sent by the WHO and UNICEF and public health authorities in most countries is that breastfeeding is the optimal way to feed all infants. There is a difference between high-income and low-income countries regarding the health consequences of not breastfeeding due, for example, to water scarcity. However, the promotion of breastfeeding is prevalent in both high- and low-income societies due to the health benefits for babies and mothers. Furthermore, breastfeeding is presented as a risk-free option that will protect and promote both the baby's and the mother's health and that will facilitate bonding between the baby and mother. Bottle-feeding is described as the risky option, and choosing to bottle-feed one's baby means refraining from doing what is natural and safe, thereby risking the baby's health (Murphy, 2000; Wolf, 2007). The scientific evidence is not as clear as one may think (Wolf, 2007; 2011). Furthermore, the message is highly value-laden, and information on risks is intertwined with notions of good motherhood (Blum, 1999; Knaak, 2010; Kukla, 2006;

Law 2000; Nihlén Fahlquist and Roeser, 2011) Furthermore, there is a group of women who have physiological, emotional and psychological problems with breastfeeding (Kelleher, 2006). Against this background, it is crucial to analyse the effects of the current message on those mothers who cannot and/or do not breastfeed. Against the background of the value-laden message, the mix of science and values in the message and the fact that we know that some groups of women have physical, emotional and psychological problems with breastfeeding, it is important to analyse the effects of this policy.

I conducted a study in order to find out more about the experiences of some of these women. The aim of the study was to gain knowledge about how non-breastfeeding mothers experience breastfeeding and the discontinuation of breastfeeding. Furthermore, the aim was to understand how these women experienced the advice and information provided by healthcare staff and official information in brochures and websites. The study was narrative and qualitative, aimed at understanding the individual situation of the respondents, that is, not to quantify them. To collect information about the experiences of this group of women, an online survey with open-ended questions was published online. An anonymous, web-based, qualitative survey was conducted, exploring the narratives of respondents who were chosen through a purposeful sample. There are obvious limitations to the study. It is not representative of mothers or of non-breastfeeding mothers. It could be the case that the mothers who do not care what authorities think about their choices did not opt to participate. In order to generalise, more extensive studies would have to be conducted. However, the point of this study is not to generalise but to explore the individual narratives of the responding mothers. The sample was purposeful, that is, only aimed at those mothers who could not and/or did not breastfeed. The only criterion for participating was that they could not or did not breastfeed their baby. Data were collected in 2012–2013 through an online survey tool called Thesistools.com. Two surveys entitled "Emotional experiences of non-breastfeeding mothers" were published on Thesistools.com. The only difference between the two surveys was that the first survey did not ask in which country the respondent was based when having babies, while the second one included that question. In the survey, the respondents were asked questions concerning their experiences of trying to breastfeed and stopping to breastfeed. They were also asked about the information that was provided about breastfeeding and bottle-feeding and how the verbal and written information made them feel. A link to the survey was published on a Swedish Facebook group for non-breastfeeding mothers called Rätten att välja-amma/flaska (The right to choose breastfeeding/bottle) and a UK-based website for mothers, Mumsnet. It was also provided to a paediatric nurse and to colleagues in the Netherlands for distribution among Dutch non-breastfeeding mothers. The responses in Swedish and Dutch have been translated by the author and a colleague. Although it is not always easy to differentiate between different causes of emotions when someone is in such a vulnerable situation as after a delivery, the survey distinguished between questions concerning emotions due to breastfeeding, the discontinuation of breastfeeding, and the advice and information provided. The online tool for creating online surveys automatically anonymised the responses. All respondents were asked the question "Do you approve of being quoted in scientific articles anonymously? Yes/No". All respondents answered yes to that question. In Sweden, there is no requirement to request ethical approval for this kind of study. The study conforms to research ethics guidelines.

Results

The five respondents to the first version of the survey are all based in Sweden, because the link to this survey was only published in Sweden. Four out of five replied in Swedish and one replied in English. The second survey got 42 responses, of which 24 stated that they gave birth and received their information in Sweden. Seven respondents stated that they gave birth and received information in the United Kingdom. Four women stated that they gave birth in the Netherlands. Three of these women received their information in the Netherlands and one received her information in Belgium. Only one respondent in the second survey did not provide details concerning which country she gave birth in, but she replied in Swedish.

All but four women tried to breastfeed at first because they wanted to and/or because it was expected of them. Almost all of the mothers had received information about infant feeding verbally from midwives and doctors as well as written information in booklets and on websites. Most women felt that the information they received strongly favoured breastfeeding. After having studied the responses written by the women, a number of themes arose, the most salient being the following:

- Depression, anxiety and pain.
- Feeling like a failure as a mother and woman.
- Loss of freedom/feeling trapped.
- Relief and guilt.

A qualitative content analysis was made. The above four themes were shared by many non-breastfeeding mothers, and they reoccurred in responses to different questions. The emotions involved appear to be related both to how they felt when trying and after trying to breastfeed and to the way in which the norm of breastfeeding was communicated to them by healthcare staff and official information. The themes are discussed in the following section.

Depression, anxiety and pain – feeling like a failed mother

When asked how they felt while trying to breastfeed, many women explained that they experienced mixed emotions. Several respondents stated that they had negative emotions because they were in pain and because they were surprised and overwhelmed by the difficulties. At the same time, many women said they wanted to keep trying despite the problems because they were told it was the best option. The most salient theme involves emotions related to depression, sadness and anxiety. Most respondents have negative emotional experiences of trying to breastfeed, discontinuing breastfeeding and of receiving advice from healthcare staff and official information on infant feeding. Many women describe that they felt pressured to continue and that they felt "like a failure" or "like a bad mother". They also use words like "inadequate" and "deviant", as well as feeling like they let their child down, probably risking their baby's health. Others state that they did not have enough milk and that they had been "connected to machines" at the hospital to try to get them to produce milk. Throughout this experience, they were continuously told that breastfeeding is the best thing. One of these women describes breastfeeding as "a nightmare" and the most important cause of her postnatal depression. Another respondent felt "heartbroken" because of her supply problems and states that she was told that everyone could breastfeed if they only tried. One of the most vivid emotional stories is the following:

> I felt awful, my daughter was crying, she didn't eat enough, lost weight, I panicked all the time and didn't know what to do. The child health center told me the problem was mine, I did something wrong, that is why she did not gain weight. I had to breastfeed even more and weigh my daughter several times a week. During this time I got mastitis in one breast and was advised to breastfeed like crazy even though I was constantly crying due to the pain. No one saw me, no one helped me, and everyone was just nagging about how good it is to breastfeed.

Relief and guilt

In response to the questions relating to the discontinuation of breastfeeding, many women describe how they harboured both positive and negative emotions. They felt relieved, but also very guilty. As one mother describes it, "When I decided to quit it was with sadness in my heart, but at the end it was the best for my son". Several women were worried that they might hurt their babies when bottle-feeding them and they felt guilty. The feelings of guilt were exacerbated by the way the breastfeeding norm was communicated by healthcare staff. They describe how they were told or made to feel like breastfeeding is the best or only option.

They felt questioned and forced to continue, and one woman writes that she hesitated to go to the public pre-school because she was ashamed. As one mother explains it, "A bad mother is what I felt like, because you *should* breastfeed, that is just how it is". Another respondent worried, "Each time we gave our baby formula I thought 'What if she gets ill, fat, etc.' ". One respondent describes how she felt guilty not merely because she stopped breastfeeding but because she did not enjoy breastfeeding. Arguably, there is a double infusion of guilt here: some women feel guilty both for not breastfeeding and for not enjoying breastfeeding. They appear to think that they are not allowed to feel good and that they ought to be ashamed of themselves for not breastfeeding. For many women it was an extremely difficult decision to stop breastfeeding. One woman describes it as the hardest decision in her life, that it hurt so much that it still makes her cry. Another woman describes it as difficult because she questioned herself as a "real mother" and she worried that her child might get allergies. But after a few months, she started to feel better because her relationship with the baby improved and she could finally enjoy her maternity leave once she stopped. One respondent describes a situation in which she was made to feel guilty by healthcare staff:

> One of the worst things I experienced actually happened in a group meeting for new parents at the local child health center. The new mothers were supposed to sit in a ring with our two-month-old babies and tell the others about ourselves, about the delivery and breastfeeding. "Well, that's nice to have to 'out' oneself like that", I thought, but I decided not to say anything about breastfeeding when it was my turn. Of course, the lady asked me about breastfeeding and I say we don't breastfeed. I added quickly that "it didn't work for us". The other mothers had told us about their struggle with breastfeeding and mastitis and God knows what but they "kept struggling" and certainly "didn't give up". Unlike me then. Then, the lady in charge turns to me (some midwife), tilted her head and says to me in front of everyone: "Well, you have heard how everyone in here keeps struggling with the feeding and I'm tooootally convinced that you would have succeeded breastfeeding if you would have kept struggling." I was completely stunned. And didn't say anything. And she keeps asking in front of everyone if I had enough support from my husband or if *he* was the one who was negative? I was so angry, but at the same time it felt wrong to start arguing about it then and there. But I regret that like hell today, honestly. Had it been today, I would have really told her off. And in the middle of all this, *this* is the saddest part, that you are so exposed when you need the most support and encouragement.

Some women had a supportive and caring midwife or paediatric nurse, but many respondents explain that instead they received inadequate information. One mother states, "When no one talks about formula, and the paediatric nurse says that she cannot 'promote' formula, you feel like a criminal, like you are doing something illegal". Another respondent says her informed choice was not respected because the midwife called her up trying to convince her to breastfeed her next child and asking whether she was "sure" she wanted to stop.

Bonding problems

Some women suffered from postnatal depression and/or had difficulties bonding with their baby, and some felt rejected by their child or started to associate the baby's being hungry and crying with physical pain. For these women, the problems associated with breastfeeding made the situation worse.

Several women describe that their relationship with the baby was impeded by the breastfeeding problems or by the way they were treated by midwives and nurses. One woman describes how she felt that healthcare staff were solely interested in the baby's well-being and did not care about *her* well-being or the fact that she experienced a crisis (postnatal depression). She also describes her bonding with the baby as severely impeded by her problems with breastfeeding, and she now regrets not having stopped earlier because the bonding got so much better once she stopped breastfeeding. One respondent writes that she often got angry when the baby was hungry because that meant she "had" to breastfeed. When she stopped, she felt like a bad woman and mother, she says. Another respondent describes how she pretended that her baby was not hungry despite crying "and didn't start to feed him until he was hysterical".

Feeling trapped – losing one's freedom

Some women did not enjoy breastfeeding at all. One woman describes how she felt when trying to breastfeed as follows: "Trapped, disgusted, in pain, anxiety/angst". Several respondents describe that they felt like they lost their freedom and felt trapped when breastfeeding. One mother states that she felt like a freer and happier person when she stopped. In addition to feeling trapped by breastfeeding, some women are burdened by the advice given and the way in which it is provided. One woman writes that she stopped believing the midwives or health visitors after she decided to stop breastfeeding and took control of the situation.

Discussion

This study indicates that some mothers who have problems breastfeeding are severely affected, become depressed and feel guilty because they stop breastfeeding. It is of course difficult to disentangle the causes and know what is due to their own situation and personality and what is due to the way the risks and benefits are being communicated by healthcare staff and by other people around them. However, the majority of the respondents felt that they were told that breastfeeding was the best option and that they did not receive adequate information and/or support about bottle-feeding. This appears to have led to a lack of self-esteem for this group of new mothers. This is the opposite of what should reasonably be expected from midwives, paediatric nurses, doctors and even the authorities informing new parents. From a normative perspective, institutions and people providing information to individuals who have recently undergone the life-changing experience of becoming a parent should be based on care and empathy. Parents should be supported in their new role. Although most healthcare providers would agree, this study illustrates the unintentional effects of such care. Arguably, the norm to breastfeed infants should not take precedence over care and empathy. Ethicists of care argue that attention to the needs of others and responding to those needs are fundamental to morality (Held, 2006; Noddings, 2002). There are good reasons to claim that this standard should be intrinsic to the care provided to new parents and families. However, the experiences shared by the respondents indicate that in many cases this is not the reality.

The goal in many countries is to increase or maintain breastfeeding numbers. There is a risk that the experiences of non-breastfeeding women are not taken seriously if infant feeding is solely analysed from a quantitative perspective. The experiences of this group of women are ethically important and should inform policy. Current information concerning infant feeding is essentially a form of risk communication in which new parents are informed about the risks and benefits of breastfeeding and bottle-feeding. Just as in the vaccination context, it is reasonable to demand that risk communication be examined on all three levels: message, procedure for communicating the message, and the message's effects. In the following section, the study is discussed from the perspective of this notion of responsible risk communication. In addition, insights from the ethics of care are used to shed light on the need to improve information about infant feeding.

Message based on individual risk-weighing

Infant feeding information essentially addresses the communication of risks and benefits about breastfeeding and bottle-feeding. That information is based on a calculation regarding such risks and benefits. There are generally two risk-weighing principles. In clinical practice, risks and benefits are commonly weighed in relation to one and the same individual, that is, whether a particular individual should be advised to take a certain medication depends on the risk–benefit ratio for that individual. In contrast, in large-scale societal projects, a collectivist risk-weighing principle is used whereby risks and benefits are weighed collectively. If the sum of all benefits outweighs the sum of all risks, the option is chosen (Hansson, 2004). Public health is a peculiar hybrid between the risk-weighing principles.

Risks and benefits are calculated collectively, but health is ultimately and fundamentally an intimate personal matter.

Public health is largely utilitarian, aimed at maximising utility, that is, the health of the population. As a utilitarian project, it is only concerned with population-based effects in terms of collective risks and benefits. Infant feeding information is based on the public health policy to promote breastfeeding because of its benefits. When midwives and nurses are told to inform individual mothers of the benefits and risks associated with infant feeding, they are not encouraged to pay attention to individual contextual aspects. The numbers indicating that the risks of bottle-feeding are greater than the risks of breastfeeding (although there is a selection regarding which risks and benefits should be included in the calculation) are used to convince individual mothers and fathers that they should choose breastfeeding (Stuebe, 2009).

The experiences shared by the non-breastfeeding mothers responding to the survey illustrate the ethical problems associated with public health policy. Despite the good intentions of public health, there are ethically problematic consequences for a group of women. How to feed one's baby is a highly intimate and complex decision, which should be made individually. There are no neutral ways to present risks, and the framing of risk messages affects how people react (Kahneman, 2011; Thaler and Sunstein, 2008). The experiences of the respondents illustrate the need to frame infant feeding in an empathetic, caring and context-sensitive way.

One-way information should be replaced by dialogue

In addition to an empathetic message, attention should be paid to the procedure, that is, to how information is being communicated. The flow of advice is currently seen as a one-way transmission of information. The underlying principle is that authorities and medical staff should tell people what the best option is. Instead, the delicate situation new parents are in suggests that a dialogue should be initiated during which new parents are encouraged to express their thoughts, emotions and concerns.

Public health campaigns sometimes fail to respect individual autonomy. There is an interesting tension between medicine, in which individual autonomy is the core value, and public health, which essentially represents a challenge to the same value. A compromise would at least include attentive dialogue instead of only authoritative one-way recommendations and prohibitions. The potentially paternalistic nature of public health and the permissibility of persuasive campaigns have been intensely debated (Buchanan, 2008; Rossi and Yudell, 2012). Although an interesting question, paternalism is not the issue here. The point made here is that communication should be attentive, empathetic and based on a dialogue. We can see that many non-breastfeeding mothers do not feel seen or heard, that they feel sad, depressed and angry.

Many women felt like failed mothers who were not doing the best for her baby or were even causing harm to their babies. These negative emotions are probably caused by a complex combination of wanting but failing to breastfeed, norms concerning motherhood, and the pressure of information from healthcare providers. The way these women feel obviously affects their sense of trust towards healthcare institutions. As one mother describes it, she stopped believing the midwives or health visitors when she decided to stop breastfeeding and took control of the situation.

Whether the major cause of these women's negative emotions stems from the fact that they cannot breastfeed or the way they are being treated by healthcare providers does not matter substantially. Even if unsuccessful breastfeeding were the major cause, information and care should be empathetic, attentive and understanding.

Ethical evaluation of effects

If breastfeeding is seen as a way to maximise public health, the effects of public health information are measured statistically in terms of how many mothers breastfeed their infants. The experiences shared by the survey respondents indicate that evaluation of infant feeding communication should also take other effects into account. It is useful to analyse the effects of breastfeeding information against the background of the ethics of care.

The ethics of care emphasises the relational aspect of moral life, which dictates that the standard of care is not being met if the person being cared for does not experience the treatment as care (Held, 2006; Noddings, 2002). If communication between healthcare providers and new families is seen as relational, the needs and emotions of the latter come to the fore. This requires sensitivity to the specific situation, the specific individual and her needs. Furthermore, the health and well-being of babies are obviously very closely connected to the health and well-being of mothers. Clearly, we cannot ask infants whether they feel cared for by the current policy. However, mothers and fathers could be invited to a dialogue. Even if the majority feel cared for by the current policy, from an ethics of care perspective, the experience of the minority should be taken into account. In another medical context, end-of-life care, the notion of "compassionate truth-telling" has been suggested as a way to talk to end-of-life patients about dying (Gwyneth, 2008). Although a different context, there are good reasons to adopt the notion of compassionate truth-telling when communicating with vulnerable people like new mothers (and fathers).

Conclusion

Risk communication is at the core of contemporary public debate. As citizens, we hear about risks every day through the media and from government authorities. The examples provided above illustrate ethical problems and challenges in terms of communicating risks. It is clear that government agencies need to actively take responsibility in order to make sure that all ethical aspects of risk communication are considered. As I have argued in this part of the book, I see three core requirements for responsibility as a virtue: care, moral imagination and practical wisdom. Arguably, risk communicators should cultivate these character traits, just like engineers ideally should. They are all involved in risk discourse, by creating, reducing, managing and communicating risks. It is not far-fetched to suggest that actively caring about potential victims of a certain policy, such as non-breastfeeding mothers or children suffering from side effects of the H1N1 vaccine, would improve policy from an ethical perspective. If risk communicators care, develop and use their moral imagination and practical wisdom, they are likely to increase the legitimacy and ethical justifiability of risk policies and risk communication. A first step is probably to start acknowledging the notion that risk communication is not merely about efficacy and efficiency, but also about ethics. The three-level framework does not offer answers, but it does provide a starting point for approaching the ethical analysis of risk communication policies. The context of nuclear energy also illuminates the need to have an open and respectful dialogue. Trust should be seen as the effect of responsible and respectful risk communication and not merely as a tool for achieving compliance. The vaccination example, as well as the infant feeding example, illustrate the importance of analysing the message. One lesson to be learnt from this analysis is simply that government agencies and healthcare representatives should be careful when communicating risks and benefits of diseases and drugs and vaccinations. People's concerns should be taken seriously. This does *not* imply a relativistic view where all statements concerning vaccines are seen as equally valid. In contemporary public debate we see, more clearly than ever, the need to distinguish between facts, opinions and values. Taking people's concerns seriously does not mean equating them, but it does mean that it cannot be taken for granted that experts know exactly what the concerns are unless attentive dialogue is established. Finally, the infant feeding, as well as the vaccination case, illuminate the complexity of public health communication. Even though intentions are good, the effects can be problematic.

Notes

1 This section of the chapter is based on Nihlén Fahlquist, J. and Roeser, S. 2015. Nuclear energy, responsible risk communication and moral emotions: a three level framework. *Journal of Risk Research* 18(3): 333–346.

2 This section of the chapter is based on Nihlén Fahlquist, J. and Roeser, S. 2015. Nuclear energy, responsible risk communication and moral emotions: a three level framework. *Journal of Risk Research* 18(3): 333–346.

3 This section is based on Nihlén Fahlquist, J. 2018. Vaccine hesitancy and trust. Ethical aspects of risk communication. *Scandinavian Journal of Public Health* 46(2): 182–188.

4 This section is based on the following articles: Nihlén Fahlquist. 2016. Experience of non-breastfeeding mothers, norms and ethically responsible risk communication. *Nursing Ethics* 23(2): 231–241 and Nihlén Fahlquist, J. and Roeser, S. 2011. Ethical problems with information on infant feeding in developed countries. *Public Health Ethics* 4(2): 192–202.

References

Asveld, L. and Roeser, S. (eds). 2009. *The Ethics of Technological Risk*. London: Earthscan.

Bick, D.E., MacArthur, C. and Lancashire, R.J. 1998. What influences the uptake and early cessation of breast feeding? *Midwifery* 14(4): 242–247.

Bijker, W., Bal, R. and Hendriks, R. 2009. *The Paradox of Scientific Authority: The Role of Scientific Advice in Democracies*. Cambridge, MA: MIT Press.

Blum L.M. 1999. *At the Breast: Ideologies of Breastfeeding and Motherhood in the Contemporary United States*. Boston, MA: Beacon.

Blume, S. 2006. Anti-vaccination movements and their interpretations. *Social Science & Medicine* 62: 628–642.

Buchanan, D.R. 2008. Autonomy, paternalism, and justice: ethical priorities in public health. *American Journal of Public Health* 98(1): 15–21.

Buck, R. and Davis, W.A. 2010. Marketing risk: emotional appeals can promote the mindless acceptance of risk. In S. Roeser (ed.), *Emotions and Risky Technologies* (pp. 61–80). Berlin: Springer.

Cooke, M., Schmeid, V. and Sheehan, A. 2007. An exploration of the relationship between post-natal distress and maternal role attainment, breastfeeding problems and breastfeeding cessation in Australia. *Midwifery* 23(1): 66–76.

Dawson, A. 2009. Herd protection as a public good: vaccination and our obligation to others. In A. Dawson and M. Verweij (eds), *Ethics, Prevention and Public Health*. Oxford: Oxford University Press.

Downs, J.S. 2011. Evaluations. In B. Fischoff, N.T. Brewer and J.S. Downs (eds), *Communicating Risks and Benefits: An Evidence-Based User's Guide* (Chapter 3). New Hampshire: US Food and Drug Administration (FDA), US Department of Health and Human Services.

Dubé, E., Laberge, C., Guay, M., Bramadat, P., Roy, R. and Bettinger, J.A. 2013. Vaccine hesitancy: an overview. *Human Vaccines & Immunotherapeutics* 9(8): 1763–1773. DOI: 10.4161/hv.24657.

Dubé, E., Gagnon, D., Ouakki, M., Bettinger, J., Guay, M., Halperin, S., Wilson, K., Graham, J., Witteman, H.O., MacDonald, S., Fisher, W., Monnais, L., Tran, D., Gagneur, A., Guichon, J., Saini, V., Heffernan, J.M., Meyer, S., Driedger, S.M., Greenberg, J. and MacDougall, H. 2016. Understanding vaccine hesitancy in Canada: results of a consultation study by the Canadian Immunization Research Network. *PLOS ONE* 11(6): 1–16.

Edholm, L. and Starbrink, A. 2017. Nej till förskoleplats för ovaccinerade? Dagens samhälle, debatt. 17 March.

European Commission. 2017. Joint statement on the occasion of the European Immunisation Week by European Commissioner for Health and Food Safety

Vytenis Andriukaitis and WHO Regional Director for Europe Zsuzsanna Jakab http://europa.eu/rapid/press-release_STATEMENT-17-1081_en.htm 24 April. (accessed 30 May 2017).

European Centre for Disease Prevention and Control (ECDC). 2017 (10 March). Epidemiological update: ongoing measles outbreak in Romania. http://ecdc.europa.eu/en/press/news/_layouts/forms/News_DispForm.aspx?ID=1559&List=8db7286c-fe2d-476c-9133-18ff4cb1b568&Source=http%3A%2F%2Fecdc%2Eeuropa%2Eeu%2Fen%2FPages%2Fhome%2Easpx (accessed 24 March 2017).

Figueroa, P.M. 2013. Risk communication surrounding the Fukushima nuclear disaster: an anthropological approach. *Asia Europe Journal* 11: 53–64.

Fischoff, B., Brewer, N.T. and Downes, J.S. 2011. *Communicating Risks and Benefits: An Evidence-Based User's Guide*. New Hampshire: US Food and Drug Administration (FDA), US Department of Health and Human Services.

Flanagan, M., Howe, D. and Nissenbaum, H. 2005. Values at play: design tradeoffs in socially-oriented game design. *Proceedings of CHI*: 751–760.

Friedman, B. 2004. Value sensitive design. In W.S. Bainbridge (ed.), *Berkshire Encyclopedia of Human–Computer Interaction*. Great Barrington: Berkshire Publishing Group.

Gustavsson, A. 2015. Gör det obligatoriskt för föräldrar att vaccinera barn. ETC ledare. www.etc.se/ledare/gor-det-obligatoriskt-foraldrar-att-vaccinera-barn 22 March 2015.

Guttman, N. and Salmon, C.T. 2004. Guilt, fear, stigma and knowledge gaps: ethical issues in public health communication. *Bioethics* 18(6): 331–352.

Gwyneth L. 2008. Withholding and withdrawing treatment: practical applications of ethical principles in end-of-life care. *South African Journal of Bioethical Law* 1(1): 24–26.

Hansson, S.O. 1996. Decision making under great uncertainty. *Philosophy of the Social Sciences* 26: 369–386.

Hansson, S.O. 2003. Are natural risks less dangerous than technological risks? *Philosophia Naturalis* 40: 43–54.

Hansson, S.O. 2004. Philosophical perspectives on risk. *Techné* 8: 10–35.

Hayenhjelm, M. 2006. Asymmetries in risk communication. *Risk Management* 8(1): 1–15.

Held V. 2006. *The Ethics of Care: Personal, Political and Global*. Oxford: Oxford University Press, 2006.

Hough-Telford, C., Kimberlin, D.W., Aban, I., Hitchcock, W.P., Almquist, J., Kratz, R. and O'Connor, A.G. 2016. Vaccine delays, refusals, and patient dismissals: a survey of pediatricians. *Pediatrics* 138(3): 1–9.

Hulme M. 2009. *Why We Disagree About Climate Change*. Cambridge: Cambridge University Press.

Ionescu, T.B. 2012. Communicating in Germany about the Fukushima accident: how direct encounter beat media representations. *Environmental Communication* 6(2): 260–267.

Jagtman, H.M. 2010. Cell broadcast trials in The Netherlands: using mobile phone technology for citizens' alarming. *Reliability Engineering & System Safety* 95(1): 18–28.

Jamieson, I.A. and Briggs, D.J. 2009. Towards effective risk discourse: the role of stakeholder partnerships. *International Journal of Risk Assessment and Management* 13(3–4): 276–292.

Kahneman, D. 2011. *Thinking, Fast and Slow*. New York: Farrar, Straus and Giroux.

Kelleher, C.M. 2006. The physical challenges of early breastfeeding. *Social Science and Medicine* 63(10): 2727–2738.

Knaak, S. 2010. Contextualising risk, constructing choice: breastfeeding and good mothering in risk society. *Health, Risk and Society* 12(4): 345–355.

Krimsky, S. and Golding, D. (eds). 1992. *Social Theories of Risk*. Westport, CT: Praeger.

Kukla, R. 2006. Ethics and ideology in breastfeeding advocacy campaigns. *Hypatia* 21: 157–181.

Kukla, R. 2008. Measuring mothering. *International Journal of Feminist Approaches to Bioethics* 1(1): 67–90.

Lanting, C. and Van Wouwe, J.P. 2007a. Peiling Melkvoeding van Zuigelingen. Borstvoeding in Nederland en relatie met certificering door stichting Zorg voor Borstvoeding. TNO-rapport KvL?

Lanting, C. and Van Wouwe, J.P. 2007b. Reasons underlying breastfeeding initiation and duration [Redenen en motieven om te starten en te stoppen met borstvoeding]. Leiden: TNO Kwaliteit van Leven.]

Larson, H., Cooper, L.Z., Eskola, J., Katz, S.L. and Ratzan, S. 2011. Addressing the vaccine confidence gap. *The Lancet* 378: 526–535.

Larsson, Å. 2015. Nej, ditt barn kan inte få autism av vaccin – här är sju myter om vaccination Metro 18 March.

Law, J. 2000. The politics of breastfeeding: assessing risk, dividing labor. *Signs: Journal of Women in Culture and Society* 25: 407–450.

Leiss, A. and Powell, D. 2004. *Mad Cows and Mother's Milk: The Perils of Poor Risk Communication*. Montreal and Kingston: McGill-Queen's University Press.

LFF (Läkemedelsförsäkringen). 2017. http://lff.se/om-lakemedelsforsakringen/ (accessed 30 May 2017).

Lofstedt, R., Bouder, F., Wardman, J. and Chakraborty, S. 2010. The changing nature of communication and regulation of risk in Europe. *Journal of Risk Research* 14(4): 409–429.

Manders-Huits, N. 2011. What values in design? The challenge of incorporating moral values into design. *Science and Engineering Ethics* 17: 271–287.

Möller, N. 2012. The concepts of risk and safety. In: S. Roeser, R. Hillerbrand, P. Sandin and M. Peterson (eds), *Handbook of Risk Theory: Epistemology, Decision Theory, Ethics and Social Implications of Risk* (pp. 55–85). Springer, Dordrecht.

Moreno, M.A., Furtner, F. and Rivara, F.P. 2011. Breastfeeding as obesity prevention. *Archives of Pediatric & Adolescent Medicine* 165(8): 772. DOI: 10.1001/archpediatrics.2011.140.

MPA (Swedish Medical Products Agency). 2013. Sweden, Karolinska Institute and seven healthcare regions. Registerstudie med fokus på neurologiska och immunrelat- erade sjukdomar efter vaccination med Pandemrix.

Munnichs, G. 2004. Whom to trust? Public concerns, late modern risks, and expert trustworthiness. *Journal of Agricultural and Environmental Ethics* 17: 113–130.

Murphy E. 2000. Risk, responsibility, and rhetoric in infant feeding. *Journal of Contemporary Ethnography* 29(291): 291–325.

Nihlén Fahlquist, J. and Roeser, S. 2011. Ethical problems with information on infant feeding in developed countries. *Public Health Ethics* 4(2): 192–202.

Nihlén Fahlquist, J. and Roeser, S. 2015 Nuclear energy, responsible risk communication and moral emotions: a three level framework. *Journal of Risk Research* 18(3): 333–346.

Noddings N. 2002. *Starting at Home: Caring and Social Policy*. Los Angeles, CA: University of California Press.

Persensky, J., Browde, S.A., Szabo, A., Peterson, L., Specht, E. and Wight, E. 2004. *Effective Risk Communication: The Nuclear Regulatory Commission's Guideline for External Risk Communication* (NUREG/BR-0308). Division of Systems Analysis and Regulatory Effectiveness: Office of Nuclear Regulatory Research.

Reventlow, S., Hvas, A.C. and Tulinius, C. 2001. In really great danger ... The concept of risk in general practice. *Scandinavian Journal of Primary Health Care* 19(2): 71–75.

Rikshandboken barnhälsovård. 2018. National guidelines for pediatric healthcare. www.rikshandboken-bhv.se/Texter/Amningsguide/Varfor-amma/ (accessed 17 July 2018).

Roeser, S. 2007. Ethical intuitions about risks. *Safety Science Monitor* 11: 1–30.

Roeser, S. 2012a. Moral emotions as guide to acceptable risk. In S. Roeser, R. Hillerbrand, M. Peterson and P. Sandin (eds), *Handbook of Risk Theory* (pp. 819–832). Dordrecht: Springer.

Roeser, S. 2012b. Risk communication, public engagement, and climate change: a role for emotions. *Risk Analysis* 32: 1033–1040.

Roeser, S. 2018. *Risk, Technology, and Moral Emotions*. London: Routledge.

Roeser, S. and Nihlén Fahlquist, J. 2013. Moral emotions and risk communication. In J. Arvai and L. Rivers (eds), *Effective Risk Communication*. London: Earthscan/Routledge.

Roeser, S. and Pesch, U. 2016. An emotional deliberation approach to risk. *Science, Technology, & Human Values* 41(2): 274–297.

Rossi, J. and Yudell, M. 2012. The use of persuasion in public health communication: an ethical critique. *Public Health Ethics* 5(2): 192–205.

Ruddick S. 1989. *Maternal Thinking: Towards a Politics of Peace*. London: The Women's Press.

Saint-Victor, D.S. and Omer, S.B. 2013. Vaccine refusal and the endgame: walking the last mile first. *Philosophical Transactions of the Royal Society B: Biological Sciences* 368(1623), 20120148.

Schmied, V. and Lupton, D. 2001. Blurring the boundaries: breastfeeding and maternal subjectivity. *Sociology of Health and Illness* 23(2): 234–250.

Schwartz, J.L. and Caplan, A.L. 2011. Ethics of vaccination programs. *Current Opinion in Virology* 1(4): 263–267.

Shrader-Frechette, K. 1991. *Risk and Rationality*. Berkeley, CA: University of California Press.

Shrader-Frechette, K. 2011. Fukushima, flawed epistemology, and black-swan events. *Ethics, Policy and Environment* 14(3): 267–272.

Skitka, L.J., Winquist, J. and Hutchinson, S. 2003. Are outcome fairness and outcome favorability distinguishable psychological constructs? A meta-analytic review. *Social Justice Research* 16: 309–341.

Slovic, P. 2000. *The Perception of Risk*. London: Earthscan.

Smith, P.J., Humiston, S.G., Marcuse, E.K., Zhao, Z., Dorell, C.G., Howes, C. and Hibbs, B. 2011. Parental delay or refusal of vaccine doses, childhood vaccination coverage at 24 months of age, and the Health Belief Model. *Public Health Reports* 126(Suppl 2): 135–146.

Soper, K. 1995. *What is Nature?: Culture, Politics and the Non-Human*. Oxford: Blackwell.

Sorell, T. 2009. Parental choice and expert knowledge in the debate about MMR and autism. In A. Dawson and M. Verweij (eds), *Ethics, Prevention and Public Health*. Oxford: Oxford University Press.

Stockholms Läns Landsting 1177. 2017 www.1177.se/Stockholm/Tema/Barn-och-foraldrar/Mat-somn-och-praktiska-rad/Amning-och-modersmjolksersattning/Varfor-rekommenderas-amning/

Stuebe, A. 2009. The risks of not breastfeeding for mothers and infants. *Reviews in Obstetrics & Gynecology* 2(4): 222–231.

Svensson, F. 2016. Advokat: Staten har tvingat på friska människor vaccin. *Svenska Dagbladet* 3 May (accessed 15 August 2017).

Taebi, B. 2011. The morally desirable option for nuclear power production. *Philosophy of Technology* 24: 169–192.

Thaler, R. and Sunstein, C. 2008. *Nudge: Improving Decisions about Health, Wealth, and Happiness*. New Haven, CT: Yale University Press.

Tversky, A. and Kahneman, D. 1974. Judgment under uncertainty: heuristics and biases. *Science* 185 Sep 27(4157): 1124–1131.

UNICEF UK/NHS. 2018. The baby-friendly initiative, Start4 life brochure. www.unicef.org.uk/babyfriendly/wp-content/uploads/sites/2/2010/11/otbs_leaflet.pdf

Van den Hoven, J. 2005. Design for values and values for design. *Information Age* 4: 4–7.

Van Esterik, P. 1994. Breastfeeding and feminism. *International Journal of Gynecology & Obstetrics* 47(Suppl. 1): S41–S54.

Van Esterik, P. 2011. Breastfeeding: a feminist issue. WABA Activity Sheet 4 World Alliance for Breastfeeding Action, available from: www.waba.org.my/resources/activitysheet/acsh4.htm (accessed 26 January 2011).

Verweij, M. 2013. Vaccination policies. In *The International Encyclopedia of Ethics*. Hobeken, NJ: Wiley.

WHO. 1981. International code of marketing of breast-milk substitutes, www.who.int/nutrition/publications/code_english.pdf (accessed 11 May 2018).

WHO. 2009. Infant and young child feeding: model chapter for textbooks for medical students and allied health professionals, http://whqlibdoc.who.int/publications/2009/9789241597494_eng.pdf (accessed 11 May 2018).

WHO. 2017. Measles and rubella. www.euro.who.int/en/health-topics/communicable-diseases/measles-and-rubella/measles-and-rubella (accessed 24 March 2017).

WHO/UNICEF. 1990. Innocenti declaration on the protection, promotion and support of breastfeeding, www.unicef.org/programme/breastfeeding/innocenti.htm (accessed 11 May 2018).

Wolf, J.B. 2007. Is breast really best? Risk and total motherhood in the National Breastfeeding Awareness Campaign. *Journal of Health Politics, Policy and Law* 32: 595–636.

Wolf, J.B. 2011. *Is Breast Best? Taking on the Breastfeeding Experts and the New High Stakes of Motherhood*. New York: New York University Press.

Wolfe, R.M. and Sharpe, L.K. 2002 Anti-vaccinationists past and present. *BMJ* 325: 430–432.

6 Children, risk and responsibility

Introduction

Virtues are developed over time or even a lifetime, through experience and role models. This notion of longevity is an essential difference between virtue ethics on the one hand and deontological and consequentialist theories on the other. For this reason, when discussing responsibility as a virtue, the role of children and their moral education and development should be highlighted. Parents and teachers ideally inspire and guide children to develop into individuals who take responsibility for themselves and others. Strangely, children are philosophically undertheorised as moral agents. This could be due to the fact that philosophical ethics has traditionally been developed by men who have, again traditionally, focused on what has been called the "public sphere". The public sphere has been conceptualised as the sphere where ethics and politics is relevant. Correspondingly, the "private sphere" was where women and children belonged and where ethics and politics were more or less considered irrelevant. As we will see, children have become more and more important in sociological and psychological research. Furthermore, the interests of children have become increasingly relevant in law with the development and adoption in 1989 by the UN General Assembly of the UN Convention on the Rights of the Child. When thinking about responsibility as a virtue, especially in relation to risk, safety and protection, I think it is natural to include children in the analysis. Children will be highly affected by environmental degradation and climate change, and they will have to learn how to deal with health risks like obesity. Given the importance of these contexts, in this chapter I will focus on children and technological risk. There are a number of new and emerging technologies that have implications for children. What is especially interesting in the context of emerging technologies is that they sometimes simultaneously expose children to risk and protect them from risks. Technological development is a threat as well as an opportunity for children.

Against this background, two questions are explored in this chapter. First, how is responsibility as a virtue relevant to the parent–child relationship? Second, what is the responsible way to use technology as a risk management tool in relation to children as vulnerable yet developing moral agents? In this chapter, I am not concerned with those contexts and activities that clearly expose children to risks, such as nuclear energy, road traffic or fast food and sugary drinks. Instead, two examples are provided from the context of information and communication technology (ICT) because they are both used to protect children from risks, and as we will see they also entail risks to children as developing moral agents. The examples illustrate the need to reflect ethically not only on the clear cases of risky technologies, but also on seemingly well-intended technologies. The need to reflect ethically on such well-intended technologies arises from the notion of responsibility as a virtue and how important it is to encourage children to develop into responsible adults. Even though this is important, it appears parents in modern society are too preoccupied with protecting children from immediate risks to reflect on the long-term risks associated with children who learn to expect to be protected, making them unable to take care of themselves and others. Before going into the questions and examples, we will set the stage by looking at the role of children in the area of philosophical ethics.

Ethics and children

According to virtue ethics, we develop virtues through experience, reflection and education. This insight makes it highly relevant to include a discussion of the parent–child relationship here. Parents, and adults in general, can encourage virtues and responsible behaviour in children. Additionally, parents ideally become better parents with experience, which makes responsibility as a virtue relevant not only to how children are encouraged to develop virtues, but to how parents develop virtues in this process. As we will see, when we talk about "parental responsibility", the notion is presumably based on a virtue-ethical notion of what it means to be a parent and on how an ideal parent reasons and acts (over time).

Although it is fairly uncontroversial and straightforward to claim that it is the responsibility of parents to protect children, tension arises when we also acknowledge that children need to learn to manage on their own. They need to learn to take care of themselves, but also how to manage societal risks at some point. In contemporary society, the number of technologies that can be used by parents and schools to protect children is increasing. This might not strike everyone as a problem. Parents should obviously protect their children and keep them safe. However, protection entails control and potential infringements of children's rights, autonomy and sense of responsibility. To be a moral agent is considered a prerequisite for moral responsibility, as we saw in Chapter 1. Because children are not considered to be autonomous moral agents, they are generally excluded from the philosophical discussion of responsibility (see e.g. Brennan and Noggle, 1997). In fact, the exclusion of children and mentally disabled persons has even been used as a means to define moral agency. However, moral agency begins in childhood, although, as noted by Pritchard, surprisingly few philosophical discussions concerning moral agency acknowledge this aspect (Pritchard, 1991). The lion's share of the limited philosophical discussion that exists concerning children revolves around interests and rights (Tiboris, 2014).

In the context of applied ethics, there are some discussions about children, primarily in bioethics and research ethics (see e.g. Fisher *et al.*, 2011; Friedman Ross, 2008; Lambert and Glacken, 2011; Porter *et al.*, 2012; Schwartz, 2012; Westra *et al.*, 2010). Traditionally, children have been excluded from medical research for reasons of protection, vulnerability and their presumed inability to consent. However, whether children can consent is a complex issue. Legally, they may not be able to consent because of an age limit, but of course they *can* consent and dissent and, ethically, it is up for discussion whether it is reasonable to accept their consent or dissent as equally important as an adult's consent or dissent.

Recently, the argument for excluding children from medical research has been criticised. It has been pointed out that the exclusion of children prevents the study and funding of health issues unique to children (e.g. Friedman Ross, 2008; Levine *et al.*, 2004). In addition, there is now evidence that children are capable of understanding and making decisions about participation in medical research (Swartling *et al.*, 2008). The question is where to draw the line between protection and respect for children's increasing autonomy. When children participate, or do not participate, in research, their need for protection and autonomy requires a delicate balancing act. If children are not allowed to participate for their own protection, research that could contribute to their health is prevented. Consequently, while this protects them from immediate risks, it may also expose them to future risks. Moreover, if they are allowed to participate, they are exposed to risks, but their autonomy is strengthened. As mentioned by Vitiello in relation to research with children who suffer from mental illness, "The specific challenge is in ensuring adequate protection for children while respecting their still-developing autonomy" (Vitiello, 2008). This discussion is based on children's interests, needs and rights.

Whereas most ethical theories do not address the specific challenges of children, care ethicists argue that care for the unprotected infant is the very foundation of ethics (e.g. Noddings, 2002). In order to act ethically in relation to other people, we need to analyse the position in which we all enter the world, i.e. as vulnerable infants. As we saw in the introduction to Part II, in the ethics of care approach initiated by Gilligan's *In a Different Voice* (1990), relationships and women's voices and experiences are illuminated. In addition, by implication, this ethical theory brings children to the fore (see e.g. Gilligan, 1990; Held, 2006; Noddings, 2002). These philosophers argue that children must be cared for so that they can learn both to care for others close to them and about distant others (Noddings, 2002).

The basic idea is that instead of being independent and self-sufficient, human beings enter the world as vulnerable subjects in need of care, dependent upon others to respond to their needs. This view challenges the liberal notion of individual autonomy as the right and principle that is central to the Western history of ideas and in current legislation. This idea of autonomy assumes that individuals are free and capable of making reasonable decisions. The notion of autonomy, traditionally meaning self-governance or self-determination, is central to many traditional theories in ethics and political theory and closely linked to agency and responsibility. On the contrary, the ethics of care emphasises the interdependency of life. We are not free and independent; rather life is characterised by unequal relationships, which makes us all highly dependent upon others. Recognising the relational character of moral life, the ethics of care criticises the idea that individuals are independent and autonomous in the sense of self-sufficiency (e.g. Verkerk, 2001). Autonomy could instead be seen as an evolving moral competence that can be impaired or enhanced by relationships and contexts and that exists in degrees (e.g. Meyers, 1989; Mullin, 2007; Verkerk, 2001).

At the same time, ethicists of care have acknowledged that caring for, or being responsible for, children is not merely about protection, but also entails the obligation to facilitate their growth (Ruddick, 1989). This notion is supported by empirical research. Psychological studies indicate that children whose parents provide opportunities for self-regulation and participation in decision-making are more likely to develop self-worth and to avoid depression (Smetana, Campione-Barr and Daddis, 2004). In addition, what is not adequately acknowledged when excluding children from ethical discussion is the gradual process whereby human beings become autonomous. To become autonomous requires practicing to make decisions (Kamii, 1982, 1984).

In order to counteract the tendency to exaggerate the importance of protection and control, it has been suggested that children should be seen as social actors with rights (James and Richards, 1999). This idea is also codified in the United Nations Convention on the Rights of the Child (UNICEF, 2018). However, the convention is far from universal adherence. Apart from the ethics of care, there is some discussion of children as developing moral agents and as rights-holders (e.g. Archard, 1993; Blustein, 1982; Brennan and Noggle, 1997; Matthews, 1980, 1984, 1994; O'Neill and Ruddick, 1979; Pritchard, 1991, 1996). The lack of philosophical and ethical analysis and discussion about children is highly problematic. First, children are one of the most vulnerable groups of people. They are exposed to higher degrees of risks than mentally healthy adults. Second, risks and risk management are at the core of contemporary society, and, as argued above, children need opportunities to learn how to become responsible individuals. The fact that children are both more vulnerable and increasingly autonomous creates tension when contemplating how the adult world should treat them.

This vulnerability is highly relevant in relation to environment and public health. For example, children bear the heaviest burden of environmentally caused health risks (Shrader-Frechette, 2007). Due to lower defences and higher exposure to toxins, children are at greater risk than adults (Cranor, 2013). In addition, children in contemporary society are increasingly at risk of developing lifestyle-related diseases like obesity and cardio-vascular diseases that will affect the quality and length of their lives (e.g. Goran *et al.*, 2003; Pei *et al.*, 2013). Thus, children are at increased risk due to their vulnerability, a fact that raises ethical urgent questions.

In spite of this increased vulnerability in some areas, which indicates that they need increased protection, they also need to be seen as increasingly responsible for their own lives. Otherwise, it is hard to imagine how they will be able to make good decisions when they grow up. This is crucial both from an individual and from a societal perspective. After all, children grow up to become leaders and decision-makers. In the future, they will have to deal with the societal challenges of their time. So, what does this mean for parents and adults generally, for society, if you will? We will explore this question in the following section and then move on to discuss responsibility as a virtue in relation to children as developing moral agents. As mentioned above, examples are provided form ICT technology.

Parental responsibility and the use of technology to protect children[1]

There are currently many technologies that can be used to protect children. For example, there is content-controlled software than can be used to block violent and pornographic material on the Internet. Another example is the "nanny cam", a camera used by parents to monitor the nanny taking care of their children. Sometimes the nanny is aware of the fact that she is being monitored, but in some cases it is done secretly when parents suspect their child is being abused. When analysing parental responsibility, it could be useful to begin by looking at another fairly simple technology that can be used to protect children from violent media content. The so-called V-chip is an electronic device that can be built into television sets to block programmes that are violent or otherwise deemed unsuitable to children (FCC, 2009). It has been argued that the V-chip wrongly reduces parental responsibility; in other words that the V-chip wrongly makes other actors responsible for what children watch on television. It has also been argued, on the contrary, that the V-chip is a tool for parents to exercise their responsibility. In this chapter, we are interested in the question of whether technology may redistribute responsibility, and we analyse the V-chip as an example of the potential of new technology to redistribute responsibility.

The nanny cam raises the question of whether it is right to use technological tools to find out whether one's children are being properly taken care of. What makes the case of the V-chip interesting is the involvement of several actors. The V-chip issue, and to some extent content-controlled software, raises the question of who is to be considered responsible for children and in what ways. In this sense, it is a question about responsibility distribution.

Van de Poel and I have argued that the arguments concerning responsibility in the V-chip case are based on either one or a combination of three notions of responsibility (Nihlén Fahlquist and Van de Poel, 2012). First, the argument that responsibility is redistributed (wrongly or rightly) by the V-chip can be interpreted as an argument concerning tasks and task responsibility, or what I called "distributing the oughts" in Part I. Second, it could refer to control and the related idea that the more/less control an agent has the greater/lesser is her responsibility. As we saw in Chapter 2, control is usually considered one of the conditions for responsibility and is related to the notion of freedom and voluntariness. If an agent was acting under compulsion when doing X, it is argued, she is not to be held responsible for doing X. If she, on the other hand, was in control and not coerced, she is to be held responsible for X. The V-chip could arguably be seen as a way of changing who is in control of what children watch on television. Finally, a thicker concept of parental responsibility based on virtue ethics and the ethics of care will be discussed as the third notion of responsibility underlying this debate. Interestingly, there are two opposed "responsibility arguments" about the V-chip: one concerning how the V-chip increases or enhances parental responsibility and the contrary that asserts the V-chip reduces parental responsibility.

V-chip technology

In the United States, the V-chip has been installed in all new television sets 12 inches or larger since January 1997. The V-chip basically functions as follows. Television stations broadcast a rating as part of the programme. Parents programme the V-chip by setting a threshold rating. All programmes above the rating are then blocked by the activated V-chip. The V-chip was introduced as part of President Bill Clinton's Telecommunications Act in 1996. Clinton argued that the V-chip could be a powerful force against "teen violence, teen pregnancy, teen drug use, and for both learning and entertainment" and that "[t]his is not censorship, this is parental responsibility" (Montgomery, 2007).

However, not everybody agrees with this assessment of how the V-chip affects responsibility. Michael Yamamoto has argued that

> [o]ne point that is often glaringly absent from discussion, however, is the responsibility of parents in supervising their children. Many continue to use the boob tube as a babysitter, if not a primary caregiver – and while content control mechanisms can be useful in some

circumstances, they can also serve simply to encourage the abdication of familial duties.

(Yamamoto, 2006)

He and other commentators on the Internet believe that the V-chip redistributes responsibility in a way that wrongly reduces parental responsibility (Calcagno, 1999).

As Van de Poel and I argue (2012), the V-chip debate refers to three different notions of responsibility, related to tasks, control and virtue. In this chapter, I will focus on responsibility as a virtue and the V-chip as an example of technology used as risk management tool for parents.

Parental responsibility as a virtue

The concept of responsibility as a virtue is highly relevant in the context of parenting. When we talk about "parental responsibility" we generally do not merely refer to isolated tasks or control, but to a long period of time, perhaps a lifetime or the rest of one's life with children, of guidance, inspiration and support. It addresses the very essence of parenthood, i.e. what it means to be a good or ideal parent and the norms and expectations associated to parenthood. Hence, the notion of parental responsibility extends beyond tasks and control, touching on the core ideas of ideal parenting. In the introduction to Part II, I argued that responsibility as a virtue consists of care, moral imagination and practical wisdom. From a general perspective, these sub-virtues seem highly relevant to raising a child. However, we need to specify what is particularly relevant in the context of parenting. This is what we will do in the following.

First, let us be clear about what we mean by parenthood. There is an extensive philosophical debate in bioethics, due to new reproductive technologies, concerning the issue of who is a parent. I will not go into that debate here, except to stipulate that my concept of a parent is not biological but more pluralist. There are several kinds of relationships that can be sufficient for parenthood (Bayne and Kolers, 2003). Furthermore, in the ethics of care debate, "mothering" is an important concept, a notion that is not restricted to biological mothers, or even to mothers, but to *those who perform maternal work* (e.g. Ruddick, 1989). In order to include all sorts of relevant activities, I prefer the term "parenting".

Utilitarian, Kantian and liberal theories are badly equipped for analysing the concept of parenthood because they are constructed for the "public sphere", whereas the family and children, traditionally including women, is part of the "private sphere". Claiming that, as a parent, one aims at maximising utility, even if it is family utility of the utility for children, would probably be considered highly questionable or at least strange. Someone who merely refers to duties and rights in relation to his children would also strike us as a bit odd, perhaps a bit rigid. Because virtue ethics focuses on how we act, feel and develop over time, it is the most suitable ethical theory in this context. Generally, parenthood appears to be about the virtues of parenting, in other words, what an ideal parent would be like. Should I be a tolerant parent or a strict parent? Should I be a friend or a leader? Parenthood appears to be more about what kind of a person and what kind of a parent one wants to be than it is about particular acts. Of course, particular acts are important, and most parents probably dwell on whether they did the right or wrong thing when they raised their voices to correct their children, ground them or give them candy for finishing their homework on time. However, it seems to me that these actions that we dwell on are relevant first and foremost because we dwell on what it is to be a parent and how we should be "good" parents over the entire time span that we are parents. We dwell on the particular acts because we want to develop traits like kindness, empathy, justice and so forth. Similarly, when we remember our own parents, we appear to think and talk about how they were, what *kind* of parents they were, not about isolated actions or situations.

In addition, parenting involves a great deal of improvisation. Parents constantly must deal with the unexpected, and the role of being a parent almost by nature or definition is likely to cause moral conflicts or role conflicts, such as between one's role as a parent and one's role as a professional. As we saw in previous chapters, Williams argues that responsibility as a virtue represents "a willingness to respond to a plurality of normative demands" (Williams, 2008). This seems to be a very suitable description of the responsibility of being a parent as well.

The concept of care is obviously important in the context of parenting. However, as we saw in the introduction to Part II, there are different views about the definition of care. Whereas some theorists argue that care is a virtue, others emphasise the relational aspects of care and see this feature as a differentiator between care theory and virtue ethics (Held, 2006; Noddings, 2002). If care were only a virtue, it would be more about having the disposition to act in a certain way, which highlights attitude, intention and motivation. Arguing that care is about relations illuminates the importance of assessing how the receiver experiences care. The development of the ethics of care has gone in different directions. One offspring is maternal ethics, and this theory contains some useful ideas that we can apply to the V-chip discussion. Maternal ethics, or the ethics of mothering, values the work traditionally done by women in taking care of children, and its proponents argue that this work should be seen as the foundation of ethics. We will now look at the concept of mothering, or "parenting".

Ruddick argues that there are three dimensions of maternal practice: preservation, growth and acceptability of one's children, as well as a meta-virtue that is the foundation for this work: "attentive love" (Ruddick, 1989). The requirements imposed on anyone doing maternal work, according to Ruddick, include preserving the life of the child, fostering the child's growth and shaping an acceptable child in the eyes of the social group to which the mother is identified. To be a mother (parent) is to be committed to meeting these demands through preservative love, nurturance and training (Ruddick, 1989). This description of maternal work, if applied to fathers as well, appears to be a reasonable notion of what parental responsibility amounts to:

- Preservation. Preserving the child's life does not require love, only the ability to see vulnerability and to respond to it. To achieve this aim is the constitutive maternal (or parental) act.
- Nurturance/growth. Fostering the child's emotional and intellectual growth is more historically and culturally specific.
- Social acceptability. Helping the child adhere to social norms.

Attentive love is what knits together maternal work, and it consists of a cognitive ability (attention) and a virtue (love) (Ruddick, 1989). There is one thing I would like to qualify about Ruddick's theory. It may be essential to create a socially acceptable child both for the sake of the child and for society, but it is also important to teach the child to reflect on the norms imposed by society. Although this may be difficult for many parents and will depend on the educational level of parents, it seems preferable to raise children who will be able to evaluate the norms of the social group instead of merely trying to conform to them. This is presumably important for helping children to become morally autonomous agents, and it may also help them to develop a sense of self-worth. One example could be children who are coming to terms with their homosexuality but who live in a very heteronormative society. This could also relate to norms concerning gender roles; for instance, it is arguably a parent's responsibility to teach his son that it is acceptable for boys to cry occasionally regardless of masculinity norms and whether the social group approves or not. MacIntyre argues that moral agents should be able to question the social and cultural order, not merely to accept it as it is (MacIntyre, 2002). If people had not started to question slavery, it is likely to have persisted. In order for children to become autonomous agents and to learn to take part in society, it is important that they learn to question norms. Furthermore, society needs its new citizens to grow up being able to question habits, customs and political decisions.

If we explicate Ruddick's requirements of preservation, nurturance and creating a socially acceptable child with attentive love as the meta-virtue, parental responsibility can be seen as a cluster of obligations and tasks. The notion of parental responsibility is primarily forward-looking with a long-term goal of producing a healthy, happy (in some sense) and socially productive person. Against this background, we can start to formulate a number of more specific forward-looking sub-responsibilities. The list is probably not exhaustive, but it gives an impression of more specific responsibilities that flow from Ruddick's requirements. It is also helpful for identifying which responsibilities are affected by technologies like the V-chip and which ones are not.

The virtue of attentive love identified by Ruddick appears to imply a meta-responsibility for seeing to it that the rest of the tasks are performed. This long-term responsibility is the continuous practice of caring for and taking an active interest in the child's well-being. The requirement of preservation implies the obligation to take care of the child, to defend the child and to love the child. The requirement of nurturance and growth is to a great extent about education, interpreted broadly. The specific responsibilities involved are to make sure the child develops morally and that the child is educated. The requirement of social acceptability entails the more specific responsibility of making sure the child does not harm others. And the fourth requirement of critical thinking requires a responsibility for teaching the child to think for herself and to question unreasonable norms.

Some additional specific responsibilities derive from a combination of Ruddick's requirements. For example, parents' responsibility for balancing the reduction of exposure to harm with the ideal of learning from mistakes stems partly from the requirement of nurturance and partly from the requirement of preservation. The responsibility for seeing to it that the child develops morally, learns to respect other people and abide by the law, is entailed by the requirement of social acceptability as well as the requirement of nurturance.

Some of the responsibilities entailed by parenthood are a consequence of the quality of the relationship between the parent and the child. However, there is a second relationship that is important when analysing parental responsibility: the one between parents and society. The first has been called the custodial relationship and the second the trustee relationship. According to Bayne and Kolers, parenthood inhabits the intersection between the two relationships. This means that, as a parent, one has responsibility both to the child and to society. "To be a parent is thus to attempt to maintain an equilibrium among the rights and responsibilities associated with two kinds of relationships that sometimes seem to pull in different directions" (Bayne and Kolers, 2008).

Table 6.1 summarises the different aspects of parental responsibility we have distinguished in this section. As we can see, whereas preservation and

Table 6.1 Parental responsibility: main relationships and requirements

Type of relation requirement	*Parent–child*	*Parent–society*
Attentive love (meta-responsibility)	X	
Preservation	X	(X)
Nurturance	X	(X)
Social acceptability	(X)	X
Critical thinking	X	X

nurturance primarily concern the parent–child relationship, social acceptability primarily concerns the parent–society relationship. To create a child who can think for herself is (or should be) in the interest of parents as well as society. The only requirement that is solely a parental responsibility that cannot be shared or transferred is the virtue of attentive love, which entails the long-term responsibility to see to it that the other obligations are met.

Parental responsibility, society's interest and the V-chip example

Against the background of the discussion above, our current notion of parental responsibility is that it is a virtue that entails the requirements of preservation, nurturance/growth, social acceptability and critical thinking. Due to the V-chip, some of these parental responsibilities may become shared with other actors such as the rating committee or the government. The question is whether this should be considered a problem. Given that the specific responsibilities listed in Table 6.1 should be considered as temporarily "transferable" or at least "sharable", it does not seem like a great concern. Most people probably think that parents are, by default, the ones who are primarily responsible for their own children. But when we analyse what this responsibility consists of, we can see that parents already share many of the tasks involved with others. Parents may be the sole responsible party in some rare cases, such as when children are home schooled and the family lives in an isolated environment. However, most people live in societies where teachers, grandparents, friends and neighbours take part in teaching a child about right and wrong, directly or indirectly. For example, many people are involved in teaching children to refrain from violence. Although exactly who these other people are may vary to some extent throughout a child's life, most children have other grown-ups who help them develop morally and prevent them from harming others. There is an old saying that "It takes a village to raise a child", and this can be seen as another way of saying that these responsibilities are shared. Historically and cross-culturally, children have been taken care of by many people, which appears to have many advantages, perhaps even evolutional ones (Hrdy, 2009). The modern, Western notion of exclusive mother–child, or even mother–father–child relationships appears somewhat marginal from such a historic and cross-cultural perspective (Seymour, 2013).

As van de Poel and I argue, it appears that the only parental responsibility mentioned in Table 6.1 that cannot be shared with others is the meta-responsibility to make sure the long-term goals are met. This is uncompromisingly the responsibility of the parent/parent-like person because it requires that the person knows the child very well and has a long-term perspective on the child's life. This is what Ruddick calls the meta-virtue of attentive love.

One may still wonder whether the government has a moral right to encourage the use of the V-chip by requiring the chip to be implemented in all TV sets and by requiring the rating of programmes. To analyse this, we must look at the particular risks against which the V-chip are intended to protect. These threats are (a) children becoming violent and criminal and (b) damage to children's mental health and well-being. The first threat concerns the requirement of raising a socially acceptable child, that is, the parent's primary responsibility to society. It may be seen as society's way of giving parents a tool for making sure their children become acceptable, which is primarily a societal interest. The second threat could be seen as related to the requirement of nurturance, that is, as a part of the responsibility to help the child grow, to protect her from harm and to not let fear and anxiety negatively affect her. Watching violent movies may cause fear and inhibit the child's development. The question, of course, remains how is parental responsibility, understood in this way, affected by the V-chip? Is it increased or reduced, and what aspects of parental responsibility and the relations of parents to the child and to society are affected?

The societal interest is justified to some extent, although caution is warranted in order to maintain the boundary between public and private matters. It appears reasonable to assert that society has a justified interest in preventing children from becoming violent grown-ups. Some research suggests a strong connection between children watching violent television content and violent behaviour (American Academy of Pediatrics, 2001; Huesmann, 2007). Furthermore, some studies suggest that children who watch violent programmes become fearful and anxious to a greater extent than children who do not consume violent media (Huesmann, 2007).

The trustee relationship implies that parents have a responsibility to society in addition to their responsibility to their children. Therefore, it is reasonable to conceive of the responsibility to prevent children from becoming violent as a shared responsibility between parents and society (government and industry). One way of sharing such responsibility is by sharing control. Because the government has decided that the V-chip should be installed in all TVs of a certain size, it could be argued that they have taken on their part of the responsibility related to violent television content. It is then up to the parents to decide whether and how to use the device, so they still have the ultimate control and responsibility. Of course, the crucial question of whether violent television content leads to violence remains unanswered, and this is not uncontroversial. However, society also has an interest in the mental health and well-being of children, and perhaps it is less controversial to claim that regardless of whether children become violent, there are few benefits of having them watch violent content. It appears likely that there is a risk to their well-being if they do watch a lot of violent content. In sum, the trustee relationship and the requirements for nurturing the child and for raising a socially acceptable child, and later a socially acceptable grown-up, make it reasonable for society to make an effort to reduce exposure of children to media violence.

One question that should be addressed is whether the V-chip is the right way to do that. Its intrusiveness is quite low, since parents retain ultimate control and can decide whether and how to use it. Of course, it also depends on whether people are actually using it, while theorists have argued that there are additional measures that should be considered, for example that parents should be educated about the media (American Academy of Pediatrics, 2001). However, in this context we are mainly interested in the ethical justifiability of the V-chip, and we believe that it is a tool for parents to use at their discretion as one way to take their parental responsibility to protect their children from unnecessary fear and anxiety as well as to help them become socially acceptable non-violent people. It is one way of reducing their children's exposure to detrimental media violence. However, if parents choose to, they are currently allowed to refrain from using the V-chip and to take their responsibility in other ways, for instance by not allowing television sets in their children's rooms and so forth.

The V-chip does not appear to substantially affect the ultimate long-term responsibility of parents. Some parental responsibilities that can be shared are shared through the use of the V-chip, but the only non-transferable and non-sharable responsibility for ensuring that all the other obligations are met – the meta-virtue of attentive love – does not appear to be reduced.

Responsibility as a virtue and children as developing moral agents

We have looked at a technology that does not appear to seriously threaten parental responsibility conceived as a virtue. I will now explore another technology that potentially affects children's developing sense of responsibility and moral agency. As a technology that potentially prevents children from developing their ability to take responsibility it is also, as we will see, something that affects parental responsibility.

Protecting children and reducing parental anxiety[2]

The global positioning system (GPS) can be used to track children's daily whereabouts, making it possible to constantly monitor children throughout their childhood and teens. This novelty, together with the fact that new technology can sometimes shed new light on old questions like responsibility and moral agency, makes this issue philosophically important. While monitoring adults has been discussed in both public debates and academically (e.g. Nissenbaum, 2010), the discussion about monitoring children has just begun, with some articles focusing on children's right to privacy in the school context (e.g. Davis, 2001; Hanafin et al, 2010). The practice of using GPS to monitor children is discussed from three angles: parental responsibility, children developing a sense of responsibility and children's right to privacy. There are several companies developing GPS applications for child protection and a number of products on the market. In Sweden, a number of preschools have started to use GPS as a means to keep track of children (Svenska Dagbladet, 2011). One company, Momenta, provides 100–150 Swedish preschools with GPS tags for vests (Göteborgsposten, 2011). An ad campaign during the autumn of 2011 by a company called Lociloci in one of the biggest newspapers in Sweden, *Dagens Nyheter*, encouraged parents to use GPS technology in mobile phones with the statement "Don't you want to know where your [child] is? Keep track of your [children]". In 2009, the British company Lok8u launched the Num8 wristwatch with GPS tracking technology that allows parents to receive text messages with the child's location (Johnson, 2009; Lok8u, 2012).

In addition to mobile phones and watches, there are now vests, jackets and shoes equipped with GPS technology. One of the products on the market, the Lok8u's so-called "nu.m8+" is securely fastened to a child's wrist and cannot be removed by the child, remaining attached unless deactivated by a parent (Lok8u, 2012). A company called Amber Alert GPS has a product called Amber Alert GPS V3, which alerts the parent every five minutes, has an SOS button and a "mom-friendly tracking portal" that allows parents to set zones, for example school and home, and to receive alerts when the child enters or exits those zones (Amber Alert, 2012). Furthermore, parents can check the history to see where their child has been during the day or week (Amber Alert, 2012). Companies like Garmin and Magellan have devices allegedly suitable for different ages. If, for example, you want to know where your teenager has been with the family car or how fast they were travelling, this is possible using their technology. According to kidsgpsguide.com, some American insurance companies even plan to give discounts to owners of cars equipped with GPS (Kidsgpsguide, 2012). In the United States, some schools have introduced GPS tracking as a part of an anti-truancy programme (Cargile, 2012; Santa Cruz, 2011). Several reasons are given for why parents should use this technology. The slogan found on a website marketing GPS tools used to track children is "Create safety for you and your child" (Barngps, 2012). The idea of creating safety for our children and putting ourselves at ease as parents is of course extremely alluring for risk-conscious parents. The most obvious reasons to use GPS are safety, security and anxiety relief. Websites marketing these products mention protection against kidnappers and child molesters as well as the risk of getting lost in the woods (Barngps, 2012; Kidsgpsguide, 2012). Additionally, according to Lociloci, the technology makes it possible for parents to give their children freedom. Instead of not allowing one's eight-year-old daughter to go on a bike adventure, the parent will be able to let her go if she uses the tracking device, thereby allegedly increasing her freedom. The same argument is made by Steve Salmon of the British company Lok8u, who hopes that the Num8 wristwatch will be used as a way to give children more freedom instead of promoting lazy parenting. "Only 20% of children are now allowed to go out and play. It's my profound hope that Num8 will help parents feel more comfortable about letting their children go out to play", Salmon said when interviewed by the *Guardian* in 2009 (Johnson, 2009).

Companies marketing GPS intended to protect children appeal to parents' feelings of anxiety and fear. This kind of marketing fits well into the contemporary obsession with risks and threats. Beck and Giddens famously argue that we live in a "risk society" (Beck, 2004; Giddens, 1999). Similarly, it has been suggested that we live in a "culture of fear" (Furedi, 2002). Altheide, writing about the 1999 Columbine shooting, discusses the event as part of an expanding discourse of fear and terrorism, "which may be defined as the pervasive communication, symbolic awareness, and expectation that danger and risk are central features of the physical and symbolic environment as people define and experience it in everyday life" (Altheide, 2009). Surveillance of children and marketing of devices that parents and schools can use to monitor children fits into this discourse (Marx and Steeves, 2010). For example, Lociloci appeals to an array of emotions and claims that its products can be used to create safety, to ease anxiety, and to show someone that you care about them (Lociloci, 2012). The following appeals to parents' fear are made on Kidsgps.com:

- Millions of parents all over the globe worry about their children. They want to keep them safe and secure and in a loving environment.
- Imagine trying to find out where your son or daughter has wandered off to and then feeling that sinking feeling in your stomach when you cannot find them. Or perhaps your worst fears are realised and your child is a victim of a kidnapping.

(Kidsgpsguide, 2012)

As can be seen from these quotes, appeals are made to the very basic fear and anxiety that parents feel when thinking about all the things that could happen and all the dangers they and their children face. This biologically based fear is of course what makes parents want to protect their children. However, it can be taken too far. As noted by Furedi, fear itself can sometimes become a greater problem than that which is feared. One example is that the fear of crime has become a greater concern than crime itself in many circumstances. Fear that children will be exposed to crime causes parents to not let their children play outside, which feeds and is fed by the flourishing "market in fear". Fear in contemporary society is "free-floating", and what we fear can change from one day to the other. The free-floating dynamic, Furedi argues, "is promoted by a culture that communicates hesitancy and anxiety towards uncertainty and continually anticipates the worst possible outcome". The culture of fear encourages society to approach every conceivable human experience as a risk to our safety that should be managed (Furedi, 2006). In such a context, it is hardly surprising that the issue of protecting our children receives special attention. What used to be called play is now symptomatically called "unsupervised children's activity" and treated as a risk.

In sociological research and childhood studies, the increased tendency to regulate and attempt control children's lives, possibly due to a perceived "childhood in crisis" has been extensively discussed (James and James, 2001; Valentine, 1996; Wyness, 2000). There are a number of risks and ethical issues that should be discussed in relation to GPS used to track children. First, is it part of parental responsibility to use GPS or to refrain from doing so? Second, how does the constant monitoring made available through this technology affect children's developing sense of responsibility? Third, do children have a right to privacy and, if they do, how does GPS affect that privacy? In the following, these questions will be discussed in the aforementioned order.

Parental responsibility and GPS

GPS, like other surveillance technology, can be used to monitor children to protect them from others, but also to protect them from themselves. The child is "equally vulnerable and dangerous, both unable to care for himself and yet more skilled than the adults in his life", making protection and control the dual goal of monitoring children (cf. Marx and Steeves, 2010, p. 217).

For parents who already do what they can to keep track of their children at all times, GPS will be an added tool. It potentially becomes part of the toolbox aimed for managing parental anxiety. Once it has become a natural part of everyday life in a family or in a school, there is a potential element of peer pressure. If others are doing it, I have to do it. Why shouldn't I? What will other parents, or the teachers, think if I choose to refrain (if it is even possible to opt out)? In the context of genetic information, there is a discussion about "the right not to know", for instance not knowing about one's genetic makeup to avoid serious psychological consequences (e.g. Andorno and Laurie, 2004). Similarly, it is possible that having constant access to information about one's child's whereabouts may cause psychological problems such as increased, as opposed to decreased, anxiety.[3] Some parents may not want to have this information, and the question is whether it will be possible to opt out if GPS becomes a standard tool for parents and teachers.

As discussed in the context of the V-chip, the core of parental responsibility can be understood as a virtue. The use of GPS to track children brings the potential conflict between the requirements of preservation and the requirements of nurturance to the fore: How do we protect our children without impeding their growth as individuals? A distinction should be made between gut reactions and reflective moral emotions (Roeser, 2009). One such reflective moral emotion could be the intuition that if a child is overprotected, she will not learn how to become a responsible human being who is able to take care of herself and to manage risks and challenges. Using GPS to monitor children could be seen as bringing to the fore a conflict between the gut feelings related to preservation that urge a parent to protect her child and the more reflective moral emotions associated with nurturance, such as inspiring one's child to intellectual and emotional growth. An important insight from the ethics of care is that care is a relationship where one has to consider the effect on the recipient of care to determine if it is genuine care as opposed to merely a caring attitude (Noddings, 2002).

By placing children in day-care centres and schools, parents delegate parts of their responsibility. This includes both the aspect of keeping them safe and the idea that children should learn from mistakes, so they can grow as independent human beings. If safety is our only goal as parents, we should not allow children to climb trees, explore the woods, use scissors, etc. To become independent adults on the other side of the parenting process, children need to learn how to go about in the world. Furthermore, we should think critically about what we are communicating when making children carry a device equipped with an SOS button that allows parents to be alerted when their children enter and exit zones and to know the exact history of a child's daily whereabouts. We appear to communicate that the world is not a safe place, that there are risks everywhere, and that neither the child nor its surroundings are trustworthy.

When day-care and school staff use GPS, the question may be even more difficult because of the potential double delegation: parents delegate responsibility to staff who may then delegate responsibility to technology. Imagine a preschool class that is out walking in the woods, exploring nature. One child gets lost and the technology does not work, so the child is not found until later (if, worst case scenario, at all). Who is responsible for that child? Who should have protected her and who failed – the teacher, the GPS device, the parent, or the company developing the technology?

Psychologists have studied the relationship between parental monitoring and children's adjustment and behaviour. The definition of "parental monitoring" used in such studies is often vague and there is little agreement as to what is involved (Jensen Racz and McMahon 2011). According to Jensen Racz and McMahon (2011), the term "parental monitoring" is often conflated with "parental knowledge". Noticing the conceptual conflation, Stattin and Kerr (2000), in an extensively cited work, distinguish between three different sources of parental knowledge: (a) child disclosure (children voluntarily tell their parents about their activities); (b) parental solicitation (parents ask children and/or their friends for information); and (c) parental control (parents use rules and restrictions to limit children's ability to participate in activities about which parents lack information) (Kerr, Stattin and Trost, 1999; Jensen Racz and McMahon, 2011). The study indicates that spontaneous and voluntary child disclosure is the most important source of parental knowledge. Stattin and Kerr (2000) hypothesise that the extent to which parents have listened actively since the child was younger and how warm and accepting they are affect the extent to which an adolescent is willing to disclose information about her activities (Stattin and Kerr, 2000). Not only does voluntary disclosure appear to be the most effective in parents' goal to gain knowledge about the child's activities, excessive parental solicitation and control may have negative effects on children's adjustment and behaviour as well as their well-being (Jensen Racz and McMahon, 2011; Kerr and Stattin, 2000).

This result is confirmed in a Japanese study investigating the increased use of information and communication technology (ICT) to monitor children (Nakayama, 2011). Notably, although reported crime in Japan has decreased by 40 per cent from 2002 to 2010, people perceive public safety as deteriorating. The experienced lack of public safety combined with the natural tendency of parents to worry about their children's safety has led to an increased interest in using ICT to monitor children. The Japanese study investigated the correlation of parenting style, i.e. "responsiveness" and "control", with the intention to use ICT to monitor their children. The result shows that in cases where parents and children describe their parenting style as high in control, the intention to use the technology is higher. However, children are more likely to accept being monitored if the parenting style is higher in responsiveness. As Nakayama suggests, a child may be more willing to accept being monitored when the relationship between parent and child is affectionate because it is then seen as a pure security measure. If, on the other hand, the relationship is characterised by overprotection and a controlling parenting style, the monitoring may instead be seen as an expression of that relationship and a parental tool to exercise that control (Nakayama, 2011). Studies in attachment theory have indicated that children with responsive parents are more likely to contribute to parental monitoring (Kerns *et al.*, 2001). This could possibly mean that a secure attachment leading to more disclosure would at least reduce the need to monitor children because of the perception that they expose themselves and others to risks.

These findings illuminate an interesting link between preservation and nurturance. By focusing parental efforts on nurturance, the parent may simultaneously contribute to preservation. When parents focus on encouraging one's child to become self-regulating and independent, intellectually and emotionally mature, the child is more likely to share information, thereby contributing to her own preservation.

In addition to psychological research supporting a receptive style of parenting, sociologists have noted a recent increased societal and parental concern about risks to children, risk posed by children, and the perceived need to control all aspects of their lives (e.g. James and James, 2001; Prout, 2005; Scraton, 1997; Wyness, 2000; Zeiher, 2001). Against this background, James and Richards (1999) argue that children should be seen as social actors with rights. Thus, there are good reasons to argue that contemporary relationships between adults and children should be based on respect and trust as opposed to control and regulation.

Children's sense of autonomy and responsibility

I have discussed responsibility as a virtue in relation to parenting. We will now look at children as developing moral agents who presumably need to practice responsibility as a virtue in order to become responsible adults, which is likely to require a certain measure of freedom. This freedom is decreasing in modern society, given the new technological opportunities to track, monitor and control children. The use of such technologies is normalised ("everybody does it"), and it is usually done for admirable reasons, i.e. parents want to protect children. However, if we are not careful this will become an ethically problematic risk trade-off.

In addition to the above-mentioned psychological findings supporting the idea that children should be seen as social actors, another factor concerns the process by which children gradually become moral agents and responsible persons. How do children learn to be responsible people? If we take Williams' concept of responsibility as a virtue as implying a readiness to respond to a plurality of normative demands (Williams, 2008) as a starting point, the question is the following: How could children be encouraged to be ready to respond to a plurality of normative demands?

In a study by Kakihara and others, youths with parents who had more rules engaged in less norm-breaking, but also had lower self-esteem (Kakihara *et al.*, 2010). Kakihara and others suggest that when parental control limits choice, the psychological autonomy of youths may be compromised. A loss of self-esteem and autonomy is likely to be counterproductive to developing responsibility as a virtue.

Whereas the aforementioned studies tested parental control generally, GPS and similar technology may also be examined in light of more general research concerning parental control and children's sense of autonomy and responsibility. A child whose parents use GPS to know all her whereabouts gets the message that someone else is in control and presumably also takes responsibility. If youths experience parental control as both care and a sign that the parents see the child as incompetent (Pomerantz and Eaton, 2000), their sense of trust and their developing sense of responsibility may both be diminished. Parental control needs to be adjusted over time to correspond to the youth's developmental abilities and her progress towards self-sufficiency (Eccles and Midgley 1993; Kakihara *et al.*, 2010). Children with parents who provide opportunities for self-regulation and participation in decision-making are more likely to develop self-worth and to avoid depression (Smetana *et al.*, 2004).

Of course, children are not legally responsible or liable. Nevertheless, they must be exposed to the world, to know how it feels to do praiseworthy and blameworthy things, in order to develop the character traits of a morally or socially responsible person.

Children and privacy

So far, we have discussed parental responsibility and children's sense of responsibility. GPS technology used to track children also illuminates the question of whether children have a right to privacy. Usually, the right to privacy is discussed in the context of the government tracking citizens. In this case, however, the context is somewhat different and involves several relationships: parent–child, teacher–child, parent–teacher/school and parent–industry. One interesting aspect of children and privacy is that, on the one hand, children are often seen as belonging to the private sphere by their very nature as dependents of their parents and because they are not normally considered moral agents. On the other hand, when children play in public spaces, e.g. at school playgrounds, they are located in the public sphere, which would have raised the issue of "privacy in public" (e.g. Nissenbaum, 1998, 2004) if they had been adults.

Do children have a right to privacy? Legally, they do have such a right. According to the United Nations Convention on the Rights of the Child (UNICEF, 2018), children have a right to privacy. Apart from the UN statement, children's right to privacy has not been widely discussed in spite of the current focus on privacy issues regarding adults (Hanafin *et al.*, 2010). Some discussion about children's right to privacy in the school context has recently been initiated. For example, it has been argued that if human beings have a right to privacy, then pupils who are children must also have this right (Davis, 2001).

I would argue that there is a strong case for children's right to privacy, albeit not necessarily as extensive as the adult equivalent. There are two primary reasons for granting children a right to privacy. First, although there are developmental, psychological and physiological differences between adults and children, both are human beings with needs, wants and desires, which require privacy. Children, like adults, have human rights, and it is reasonable to assert that children should have some private space. We should care and empathise with children to the extent that we take into account how they experience privacy intrusions, i.e. how they feel about being constantly monitored. In a study about the use of CCTV cameras in British schools, pupils were asked how they experienced being monitored. Most of the pupils felt that the cameras undermined privacy and represented a manifestation of mistrust (Taylor, 2010). Hence, not only do they feel bad about being monitored through technology, but it affects the trust and relationship between adults and children. As mentioned before, there has, during recent years, been an increased effort to monitor and regulate children's lives, and the circumstances under which children are free from adult control have seriously decreased (Valentine, 1996). There has been a "domestication" of children's lives whereby they are removed from the streets, and an institutionalisation whereby their activities are restricted and supervised by professionals (Prout, 2005; Zeiher, 2001).

As has been argued by sociologists and childhood scholars, we should start seeing children not as subordinate, inferior and in need of control, restriction and protection, but as "competent social actors" (Prout, 2005; Wyness, 2000). As competent social actors they, just like adults, should have a private sphere. The second reason is instrumental, i.e. it views privacy as a value because it is necessary for something else to occur. From a psychological perspective, privacy is considered an important prerequisite for the development of a self and for regulating the boundaries between self and society (Parke and Sawin, 1979).

Conclusion

The relationship between risks, responsibility and children is complex. Children are generally more exposed to risks than adults. At the same time, they are developing moral agents who need to practice in order to become responsible. In addition, parents are responsible for children, and one of their main tasks is to protect them. However, an additional task is to promote their independence in order to foster their growth. We have seen that there are new technologies that are being used to protect children. To be a parent is, essentially, to be more or less constantly preoccupied with the safety of one's child. Previously, this meant keeping an eye on the child when possible and trusting other adults such as extended family, neighbours and school staff to also watch the child. As the child grew older, parents needed to trust their child to take care of herself. Today, technology provides parents and teachers with enormously expanded opportunities to protect children.

However, the benefits of these technologies for assuaging parental anxiety should not distract us from seeing the bigger picture and the ethical nuances. There is a difference in degree when comparing the V-chip and GPS. The V-chip is an example of a relatively limited tool that arguably can be used to help exert parental responsibility both towards the child and society. Protecting the child against violent media content and reducing the risk of children acting out the violence they are exposed to do not appear to diminish the core of parental responsibility conceived as a virtue.

In contrast to the V-chip, using GPS to constantly track children could lead to the preoccupation with children's safety spinning out of control. GPS could be used to monitor young pre-schoolers when they are out in the woods, but it could also be used to track and monitor all the whereabouts of teenagers. The former probably would not be considered an infringement on the rights of children or an impediment to the development of their sense of responsibility. However, the question should still be asked, and it is important that we do not take for granted that preschoolers are not moral agents or that they do not have rights. When it comes to teenagers in general, I would argue that tracking them is indeed an infringement of their privacy and a threat to their developing a sense of responsibility. Of course, there may be exceptions, such as when a teenager is suicidal, exposed to other kinds of unusual threats, or suffers from certain medical or psychological conditions. However, in normal cases, it seems to me ethically wrong to track teenagers. I would say that even 11 and 12-year-olds should be given extensive privacy and freedom to practice responsibility as a virtue. The question of where to draw the line between protection and privacy should be negotiated between adults and children, and the power inequality between adult and child should not be misused. Parent–child and teacher–child relationships should ideally be based on trust and a gradual transfer of responsibility, something that may be hampered by using monitoring technology without careful deliberation and caution. We need to differentiate between gut feelings associated with preservation and reflective moral emotions associated with nurturance and growth. The main general problem I see at this point is that new technologies become normalised quickly, leading both parents and children to get used to them. Before that happens, the long-term consequences and ethical problems of these technologies need to be assessed and evaluated. Every new technology should be analysed in the context of how adults should ideally take responsibility for children, as well as what that technology could mean for children's developing sense of moral agency and responsibility. This is necessary to avoid harming children, and hence future decision-makers, in the long run.

Notes

1 This section is based on Nihlén Fahlquist, J. and van de Poel, I. 2012. Technology and parental responsibility: the case of the V-chip. *Science and Engineering Ethics*.
2 This section is based on Nihlén Fahlquist, J. 2015. Responsibility and Privacy – Ethical Aspects of Using GPS to Track Children. *Children & Society* 29(1): 38–47.
3 I would like to thank Peter Kroes for making this point in a colloquium discussion.

References

Altheide, D.L. 2009. The Columbine shootings and the discourse of fear. *American Behavioral Scientist* 52: 1354–1370.

Ambert Alert website. 2012. https://amberalertgps.com (accessed 21 May 2018).

American Academy of Pediatrics: Committee on Public Education. 2001. Children, adolescents, and television. *Pediatrics* 107: 423–426.

Andorno, R. and Laurie, G. 2004. Right not to know: an autonomy based approach. *Journal of Medical Ethics* 30: 435–440.

Barngps website. 2018. Available at http://barngps.se/ (accessed 11 May 2018).

Bayne, T. and Kolers, A. 2003. Toward a pluralist account on parenthood. *Bioethics* 17(3): 221–242.

Bayne, T. and Kolers, A. 2008. Parenthood and procreation. In Edward N. Zalta (ed.), *Stanford Encyclopedia of Philosophy* (Fall edition). http://plato.stanford.edu/archives/fall2008/entries/parenthood/

Beck, U. 2004. *Risk Society: Towards a New Modernity*. Sage: London.

Brennan, S. and Noggle, R. 1997. Children's rights, parents' rights and family justice. *Social Theory and Practice* 23(1): 1–26.

Calcagno, P. 1999. V-chip parenting. Ludwig von Mises Institute. http://mises.org/daily/245 (accessed 31 May 2009).

Cargile, E. 2012. AISD: GPS tracking to boost attendance. Kxan.com, 27 August.

Cranor, C.F. 2013. *Legally Poisoned: How the Law Puts Us at Risk from Toxins*. Cambridge, MA: Harvard University Press?

Davis, A. 2001. Do children have privacy rights in the classroom? *Studies in Philosophy and Education* 20: 245–254.

Eccles, J.S. and Midgley, C. 1993. Development during adolescence: the impact of stage-environment fit on young adolescents' experiences in schools and in families. *American Psychology* 48(2): 90–101.

FCC (Federal Communications Commission). 2009. The V-chip: putting restrictions on what your children. Federal Communications Commission. www.fcc.gov/cgb/consumerfacts/vchip.html (accessed 15 January 2010).

Fisher, H.R., McKevitt, C. and Boaz, A. 2011. Why do parents enrol their children in research: a narrative synthesis. *Journal of Medical Ethics: The Journal of the Institute of Medical Ethics* 37(9): 544–551.

Friedman Ross, L. 2008. *Children in Medical Research*. Oxford: Oxford University Press.

Furedi, F. 2002. *The Culture of Fear*. London: Continuum.

Furedi, F. 2006. *The Culture of Fear Revisited*. Continuum: London.

Giddens, A. 1999. Risk and responsibility. *Modern Law Review* 62: 1–10.

Gilligan, C. 1990. *In a Different Voice: Psychological Theory and Women's Development*. Cambridge, MA: Harvard University Press.

Goran M.I., Ball, G.D. and Cruz, M.L. 2003. Obesity and risk of type 2 diabetes and cardiovascular disease in children and adolescents. *Journal of Clinical Endocrinology and Metabolism* 88(4): 1417–1427.

Göteborgsposten. 2011. Allt vanligare med gps på förskolan. 21 September 2011.

Hanafin, J., Donoghue, T., Flynn, M. and Shevlin, M. 2010. The primary school's invasion of the privacy of the child: unmasking the potential of some current practices. *Educational Studies* 36: 143–152.

Held, V. 2006. *The Ethics of Care: Personal, Political and Global*. Oxford: Oxford University Press.

Hrdy, S.B. 2009. *Mothers and Others: The Evolutionary Origins of Mutual Understanding*. Cambridge, MA: Harvard University Press.

Huesmann, R. 2007. The impact of electronic media violence. Scientific theory and research. *Journal of Adolescent Health* 41: 6–13.

James, A.L. and James, A. 2001. Tightening the net: children, community and control. *British Journal of Sociology* 52: 211–228.

James, A.L. and Richards, M. 1999. Sociological perspectives, family policy, family law and children: adult thinking and sociological tinkering. *Journal of Social Welfare and Family Law* 21: 23–39.

Jensen Racz, S. and McMahon, R.J. 2011. The relationship between parental knowledge and monitoring and child and adolescent conduct problems: a 10-year update. *Clinical Child and Family Psychological Review* 14: 377–398.

Johnson, B. 2009. GPS wristwatch helps parents track children. *Guardian* 12 January.

Kakihara, F., Tilton-Weaver, L., Kerr, M. and Stattin, H. 2010. The relationship of parental control to youth adjustment: do youths' feelings about their parents play a role? *Journal of Youth Adolescence* 39: 1442–1456.

Kamii, C. 1982. *Number in Preschool and Kindergarten*. Washington, DC: NAEYC.

Kamii, C. 1984. The aim of education envisioned by Piaget. *Phi Delta Kappan* 65(6): 410–415.

Kerns, K.A., Aspelmeier, J.E., Gentzler, A.L. and Grabill, C.M. 2001. Parent–child attachment and monitoring in middle childhood. *Journal of Family Psychology* 15: 69–81.

Kerr, M. and Stattin, H. 2000. What parents know and how they know and several forms of adolescent adjustment: further support for a reinterpretation of monitoring. *Developmental Psychology* 36: 366–380.

Kerr, M., Stattin, H. and Trost, K. 1999. To know you is to trust you: parents' trust is rooted in child disclosure of information. *Journal of Adolescence* 22: 737–752.

Kidsgpsguide website. 2012. Available at www.kidsgpsguide.com (accessed 11 May 2018).

Lambert, V. and Glacken, M. 2011. Engaging with children in research: theoretical and practical implications of negotiating informed consent/assent. *Nursing Ethics: An International Journal for Health Care Professionals* 18(6): 781–801.

Lok8u website. 2018. Available at www.lok8u.com (accessed 21 May 2018).

MacIntyre, A. 2002. *Ethics and Politics: Selected Essays, Vol. 2*. Cambridge: Cambridge University Press.

Marx, G.T. and Steeves, V. 2010. From the beginning: children as subjects and agents of surveillance. *Surveillance and Society* 7: 192–230.

Meyers, D.T. 1989. *Self, Society and Personal Choice*. New York: Columbia University Press.

Montgomery, K.C. 2007. *Generation Digital: Politics, Commerce, and Childhood in the Age of the Internet*. Cambridge, MA: MIT Press.

Mullin, A. 2007. Children, autonomy, and care. *Journal of Social Philosophy* 38: 536–553.

Nakayama, M. 2011. Parenting style and parental monitoring with information communication technology: a study on Japanese junior high school students and their parents. *Computers in Human Behavior* 27: 1800–1805.

Nihlén Fahlquist, J. 2015. Responsibility and privacy – Ethical aspects of using GPS to track children. *Children & Society* 29(1): 38–47.

Nihlén Fahlquist, J. and Van de Poel, I. 2012. Technology and parental responsibility: the case of the V-chip. *Science and Engineering Ethics* DOI: 10.1007/s11948-010-9222-6.

Nissenbaum, H. 1998. Protecting privacy in an information age: the problem with privacy in public. *Law and Philosophy* 17: 559–596.

Nissenbaum, H. 2004. Privacy as contextual integrity. *Washington Law Review* 79: 119–157.

Nissenbaum, H. 2010. *Privacy in Context: Technology, Policy, and the Integrity of Social Life*. Stanford, CA: Stanford Law Books.

Noddings N. 2002. *Starting at Home: Caring and Social Policy*. Berkeley, CA: University of California Press.

Parke, R.D. and Sawin, D.B. 1979. Children's privacy in the home: developmental, ecological, and child-rearing determinants. *Environment and Behavior* 11: 87–104.

Pomerantz, E.M. and Eaton, M.M. 2000. Developmental differences in children's conceptions of parental control: "They love me, but they make me feel incompetent'. *Merrill-Palmer Quarterly* 46: 140–167.

Pritchard, M. 1991. *On Becoming Responsible*. Lawrence, KS: University Press of Kansas.

Pritchard, M.S. 1996. *Reasonable Children: Moral Education and Moral Learning*, Lawrence, KS: University Press of Kansas.

Prout, A. 2005. *The Future of Childhood*. New York: Routledge.

Roeser, S. 2009. The relation between cognition and affect in moral judgments about risk. In L. Asveld and S. Roeser (eds), *The Ethics of Technological Risk* (pp. 182–201). London: Earthscan.

Ruddick S. 1989. *Maternal Thinking*. Boston, MA: Beacon Press.

Santa Cruz, N. 2011. For chronic truants, a GPS program can help them make the grade. *Los Angeles Times* 25 February.

Shrader-Frechette, K. 2007. *Taking Action, Saving Lives: Our Duties to Protect Environmental and Health*. Oxford: Oxford University Press.

Scraton, P. 1997. *Childhood in Crisis*. London: UCL Press.

Seymour, S.C. 2013. "It takes a village to raise a child": Attachment theory and multiple child care in Alor, Indonesia, and in North India. In N. Quinn and J.M. Mageo (eds), *Attachment Reconsidered: Cultural Perspectives on a Western Theory* (Chapter 4). New York: Palgrave Macmillan.

Smetana, J.G., Campione-Barr, N. and Daddis, C. 2004. Longitudinal development of family decision-making: defining healthy behavioral autonomy for middle-class African American adolescents. *Child Development* 75: 1418–1434.

Stattin, H. and Kerr, M. 2000. Parental monitoring: a reinterpretation. *Child Development* 71: 1072–1085.

Svenska Dagbladet. 2011. Allt vanligare med GPS på förskolan. 21 September.

Swartling, U., Helgesson, G., Hansson, M.G. and Ludvigsson, J. 2008. Parental authority, research interests and children's right to decide in medical research – an uneasy tension? *Clinical Ethics* 3(2): 69–74.

Taylor, E. 2010. I spy with my little eye: the use of CCTV in schools and the impact on privacy. *Sociological Review* 58: 381–405.

Tiboris, M. 2014. Blaming the kids: children's agency and diminished responsibility. *Journal of Applied Philosophy* 31(1): 77–90.

UNICEF. 2018. United Nations Convention on the Rights of the Child (CRC). Available at www.unicef.org/crc/ (accessed 21 May 2018).

Valentine, G. 1996. Children should be seen and not heard? The role of children in public space. *Urban Geography* 17: 205–220.

Verkerk, M.A. 2001. The care perspective and autonomy. *Medicine, Health Care and Philosophy* 4(3): 289–294.

Vitiello, B. 2008. Effectively obtaining informed consent for child and adolescent participation in mental health research. *Ethics & Behaviour* 18(2–3): 182–198.

Westra, A.E., Sukhai, R.N., Wit, J.M., de Beaufort, I.D. and Cohen, A.F. 2010. Acceptable risks and burdens for children in research without direct benefit: a systematic analysis of the decisions made by the Dutch Central Committee. *Journal of Medical Ethics* 36(7): 420–424.

Williams, G. 2008. Responsibility as a virtue. *Ethical Theory and Moral Practice* 11(4): 455–470.

Wyness, M. 2000. *Contesting Childhood.* London: Routledge.

Yamamoto, M. 2006. The real V-chip villain: Bad parents. 18 April 2006 CNet News.com Newsblog http://news.cnet.com/8301-10784_3-6062289-7.html.

Zeiher, H. 2001. Children's islands in space and time: the impact of spatial differentiation on children's ways of shaping social life. In M. du Bois-Reymond, H. Sünker and H-H. Krüger (eds), *Childhood in Europe, Approaches – Trends – Findings* (pp. 139–159). New York: Peter Lang.

Index

Page numbers in **bold** denote tables.